WIRED WISDOM

WIRED WISDOM

HOW TO AGE BETTER ONLINE

Eszter Hargittai & John Palfrey

THE UNIVERSITY OF CHICAGO PRESS

Chicago and London

The University of Chicago Press, Chicago 60637
The University of Chicago Press, Ltd., London

Published 2025
Printed in the United States of America

34 33 32 31 30 29 28 27 26 25   1 2 3 4 5

ISBN-13: 978-0-226-82345-4 (cloth)
ISBN-13: 978-0-226-84139-7 (paper)
ISBN-13: 978-0-226-82346-1 (e-book)
DOI: https://doi.org/10.7208/chicago/9780226823461.001.0001

Library of Congress Cataloging-in-Publication Data

Names: Hargittai, Eszter, 1973– author. | Palfrey, John, 1972– author.
Title: Wired wisdom : how to age better online / Eszter Hargittai and John Palfrey.
Description: Chicago : The University of Chicago Press, 2025. | Includes bibliographical references and index.
Identifiers: LCCN 2024047697 | ISBN 9780226823454 (cloth) | ISBN 9780226841397 (paperback) | ISBN 9780226823461 (ebook)
Subjects: LCSH: Technology and older people. | Internet and older people. | Digital divide.
Classification: LCC HQ1061 .H3375 2025 | DDC 305.26—dc23/eng/20250115
LC record available at https://lccn.loc.gov/2024047697

♾ This paper meets the requirements of ANSI/NISO Z39.48-1992 (Permanence of Paper).

To Magdolna and István Hargittai, Judy and Sean Palfrey, and Mary and John Carter for showing us how to age successfully

CONTENTS

1. Wired Wisdom 1

2. Adoption: Are Older People Less Likely to Use New Tech? 17

3. Support: How Do Over-Sixties Seek Help? 39

4. Safety and Security: The Greater the Age, the Easier the Target? 61

5. Privacy: What's Worth the Price of Personal Data? 85

6. Misinformation: Why Do Skeptics Spread Fake News? 107

7. Well-Being: Does Tech Increase Loneliness? 131

8. Learning: Can New Tech Teach New Tricks? 153

9. Lessons—For Older Adults, Their Families, Friends, and Society 177

10. Top Ten Takeaways 189

Acknowledgments 191 *Methodological Appendix* 193

Notes 203 *Bibliography* 235 *Index* 267

1

Wired Wisdom

How can you thrive in the digital world? This single question motivates much of our research, and particularly this book on technology users aged sixty and over. If you believe many popular accounts, most older internet or computer technology users are not thriving. They are slow to adopt new technologies that could help them. They are easily scammed. They spread misinformation. They cannot compete with younger colleagues who are "born digital," with supposedly inherent abilities to adopt and wield technology.

If you are sixty or over, these popular accounts might not match your reality. You might not only adopt new technologies quickly but also teach family and friends how to use them. You might show great prowess in identifying online scams and fake news. You might work in tech, developing new tools while you watch younger colleagues struggle.

In the pages that follow, we bust the myths about clueless and hapless older adults' relationship to digital technologies. You will read many reasons to doubt the stereotypes about older technology users. If this book—and the research that informs it—shows one thing, it is diversity. The harmful stereotypes about older users' inability to use technology only adds to the problem, preventing steps that would help us all benefit from older technology users' contributions.

We offer clear advice about how older adults—defined by both the United Nations and the World Health Organization as people sixty and over[1]—can benefit from using new technologies while avoiding their pitfalls. We hope our findings are useful for all readers: older adults themselves, those who care about and for them, and policymakers, who help make essential resources—including technologies—available to all members of society.

We agree with other writers and researchers[2]—and many of our interviewees—that the years after sixty are often deeply fulfilling, fun, and meaningful on multiple dimensions. We strive through this work to help readers and their loved ones age successfully. What do we mean by aging successfully? We explain in the pages that follow.

As you browse your library, bookstore, or online bookseller, you may find many books on younger internet users, but few on older adults. Many early studies of internet use focused on younger demographics, as younger people were more likely to be online in the late 1990s and early 2000s.[3] What quickly became clear, however, was that youth and young adults varied considerably in their online skills and activities, challenging the notion of a universally savvy "born digital" generation almost as soon as the term was established.[4] Instead of age, different educational, social, and economic resources led to a digital divide in society; the more privileged were more likely to start using the internet than their less privileged counterparts.[5] We will see some of these same trends among internet users sixty and over, or "over-sixties."

The research on younger users also soon made clear that even after their adoption of digital technologies, differences remained in their uses and skills.[6] So, for example, simply ensuring that everyone had access to an internet connection would not be enough: if people lacked the skills to make the most of technologies, then they would continue to trail those who embraced technologies for increasing parts of their everyday lives. In sum: people vary in their access, skills, and uses, which in turn all influence the potential benefits they can reap from new technologies.[7]

The study of older adults and technology is an especially unsettled area of research. There are more gaps in our knowledge than there are settled questions. There are contradictions in the data, blind spots we have not yet discovered, questions we have not yet asked, geographies and cultures that have escaped researchers' gaze.

But, as you will see, there has also been progress. In writing this book, our principal job is to introduce the ideas we think are "least wrong"—where the best scientific evidence can inform a practice with a reasonable degree of certainty. It is not our intention to denigrate the terrific researchers on whose good work we draw, nor do we wallow in false modesty about our own study of this topic. But the science about technology and older adults is uncertain. Why make recommendations on topics continually changing without conclusive results? The alternative would be never giving advice at all, which would be a shame as scholarship has certainly made strides with useful takeaways that we translate for everyday cases on these pages.

We have another reason to write this book. Older adults often feel ignored and invisible, despite their making up an increasing portion of the world's population. Research documents this phenomenon across a range of countries, including Australia, Canada, New Zealand, the United Kingdom, and the United States.[8] And it is not a misguided perception. Research also shows that many younger people indeed have negative preconceptions about older adults.[9] In this book, we acknowledge the importance of this population segment to a healthy and inclusive society.

More and more older adults around the world are using new technologies.[10] This growth coincides with a few related major trends: the rapid demographic shift in the number and proportion of older persons, the rapid dissemination of information and communication technologies, and the growing importance of these technologies across many sectors of life.

Despite the fact that adults aged sixty and over represent the fastest-growing demographic segment worldwide,[11] research to understand this population has not kept up.[12] What studies do show consistently is that many older adults are more digitally connected than ever and that connectivity can have meaningful implications for their well-being. The exact nature of this relationship between technology use and health and well-being is complex and nuanced. But the pages that follow share the conclusions we can confidently draw based on the research to date.

While over-sixties are generally less likely to use technology, there is fairly dramatic growth in this age group in particular. Because many other age groups have come close to universal adoption of basic technologies such as mobile devices, growth for some technologies is now highest among older adults. For instance, the number of over-sixties who own smartphones increased from 13% in 2012 to 61% in 2021 in the United States. While this group's smartphone adoption trails other segments of the population, this statistic shows the growing importance of new technologies to older adults.[13]

Let's stick with the example of smartphone adoption as

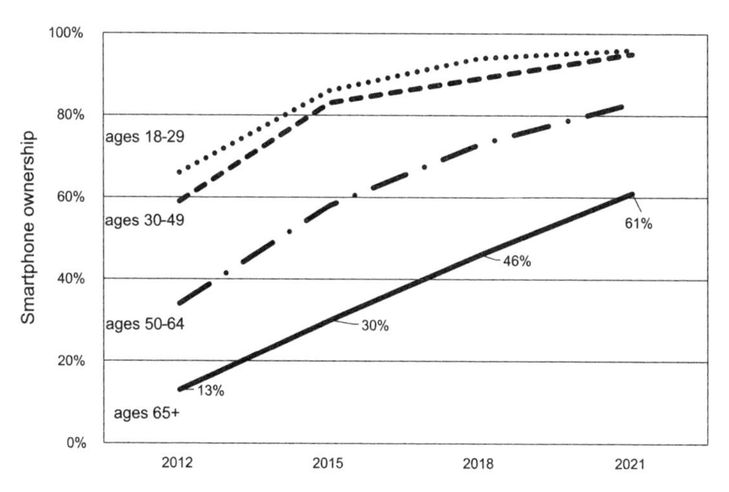

Smartphone ownership among Americans of different ages from 2012–2021. (Data source: Pew Research Center, Faverio)

one lens on the topic. There is a lot of variation in terms of adoption. Among over-sixties, the younger the person, the more likely that person owns a smartphone—in other words, a sixty-five-year-old is more likely to have a smartphone than a seventy-five-year-old, who is more likely to have a smartphone than an eighty-five-year-old. Research into digital inequality repeatedly shows that many factors beyond age, such as income and education, link to variations in technology adoption and use.[14] For example, a wealthier person is more likely to own a smartphone than a less wealthy person. Someone with a college degree is more likely to have a smartphone than someone without a college degree, and so forth.[15] While our goal is to identify generalizable patterns across older adults of differing social circumstances, it would be wrong to ignore the influences of socioeconomic differences on successful aging and technology.

Older adults use digital tools to connect to others through social media and use other technologies that facilitate social connection. As in other age groups, older adults have found that platforms such as Facebook allow them to connect to others they know today, those they have known in the past, and those who share similar interests across the globe. As we explore the connection between older adults, their technology use, and their health and well-being, we often focus on social media use and engagement in online communities.

Again, our goal is to help people "age successfully" with technology. But how do we define success? In the academic literature, the successful aging idea dates back to a theory by social scientists John Rowe and Robert Louis Kahn.[16] In 1997, these experts defined success this way:

> Successful aging is multidimensional, encompassing the avoidance of disease and disability, the maintenance of high physical and cognitive function, and sustained engagement in social and productive activities.[17]

Other scholars use different terms, such as "healthy aging," "active aging," "productive aging," and "aging well."[18] Whatever the term, we believe that focusing on how digital media can help older adults live healthy and happy lives is a worthwhile endeavor,[19] even if it first requires wading through considerable research and much misinformed public rhetoric.

We aim to describe the relationship between technology and aging that grew more than a quarter century after Rowe and Kahn's theory. Whether consciously or not, many over-sixties use new technologies as they pursue successful aging.

As already noted, older adults' experiences with technology differ. Some of these differences relate to life stage—someone in their early sixties who has an office job is in a very different spot from someone in their late eighties who was already retired when internet use became widespread. Other reasons for varied uses, however, go well beyond age and life stage.

For over two decades, our research has focused on a crucial differentiating factor: internet skills, or digital literacy. This concept refers to people's ability to use technologies effectively and efficiently.[20] At the most basic level, digital literacy concerns awareness of what is possible to do with digital media. Someone who does not know that something is possible to do is unlikely to seek it out, which is why awareness of various options is so crucial. Take, for example, smartphone notifications. These can easily get overwhelming, yet not all users are aware that they have a say in whether and how they receive notifications. The first step to changing notification settings is knowing that such an option exists. A more specific example is the Be My Eyes app, which connects blind and low-vision users to volunteers and companies to help them with visual questions through live video. When the user has a question, the app connects them to someone who can look at the object in question and offer advice. Applications include finding certain settings on a device or picking the right item out of the fridge. This free service can help only those who are aware that it exists.

Research finds that some people do not engage with digital tools (including the internet) because they think these technologies have nothing to offer them.[21] To us, it is inconceivable that there is anybody out there for whom the internet could not be helpful, whether in the realm of work or leisure, in the domain of social connection or information access, for the purposes of entertainment or intellectual enrichment. Rather, thinking that the internet is not relevant to one's situation and interests reflects a lack of skills, a lack of awareness of what is possible. To this end, spreading the word about online opportunities is one way to expand informed internet uses. These can range from taking certain security actions to changing privacy settings, from installing helpful apps to knowing how to find good deals or support networks online. Ultimately, much of successful aging with technology is about having the skills to navigate the vast digital world in beneficial ways.

As you have likely gathered from the cover, this book is coauthored. We bring different academic and professional backgrounds to our collaboration on this work. We bring different research methodologies and different experiences in life and in the practice of bringing about change.

Eszter (age fifty at the time of writing) is a sociologist and communication scholar who has conducted several academic studies on older adults' technology uses. Eszter has led groups of researchers establishing a deep understanding of how people use digital technologies and the implications of that use. For more than two decades, Eszter has researched people representing a wide range of ages and living in varied geographies; her recent work in Zurich has emphasized older adults.

John (age fifty-one at the time of writing) trained as a lawyer and, over the same period of about two decades, has studied the ways people use technologies. John has written broadly on issues related to technology and how people of multiple ages use it. John has studied and written, mostly with media scholars Urs Gasser and Sandra Cortesi, books and articles on how young people use new technologies, including *Born Digital* (with Urs Gasser). In partnership with his colleagues, John has worked to bring about changes in policy with respect to how the internet is regulated—with the goal of serving public interest, not just increasing corporate profits.

Today, John is president of the John D. and Catherine T. MacArthur Foundation. The MacArthur Foundation has funded research both in the area of aging and in the field of technology, digital media, learning, and the public interest, including work by Rowe and Kahn, mentioned earlier. The MacArthur family noted in the original articles of incorporation that the foundation should provide support "to investigate and attempt to find solutions to the social, economic, mental, and physical problems of retired persons." One of the most important investments by the MacArthur Foundation over the years was called the Research Network on Successful Aging.[22]

We met in 2007 thanks to an introduction by John Bracken, a program officer at the MacArthur Foundation at the time. We have collaborated on both research and teaching in the past. Over a decade before teaching on Zoom spread like wildfire during the COVID-19 pandemic, we cotaught a seminar for our graduate students at Harvard University and Northwestern University over videoconferencing technology. It was not a perfect course, but we learned a lot—and in part it led to a collaboration that has produced this book.

Although our research in the early 2000s was not about older adults, our work from then very much informs our writing for this book. For more than two decades, Eszter has conducted work on digital inequality and has documented across time and geographies how people's social status connects with their internet skills and online activities. This recognition that people's background characteristics such as age, gender, and disability status influence their online opportunities very much informs the work we have done on this book. Eszter's first paper about older adults appeared in 2006; since then she has published over a dozen peer-reviewed academic articles on the topic.

After publishing *Born Digital* in 2008, John went on to publish a book (among many others) about parenting in the digital age.[23] Writing about older adults is a natural extension of his work examining people's online experiences throughout the life course. Given its focus on actionable steps and policy recommendations, the book also benefits from his extensive writing about legal issues related to digital technologies. In his work at the MacArthur Foundation, he and his colleagues have built on the work that the foundation has done for decades in support of older adults and in the field of technology for the public interest. These combined commitments lie at the heart of his interest in partnering with Eszter on this book project.

Work by both of us has raised questions about how best to achieve accessible information and communication technologies, making them within reach to a broad spectrum of the population. Until recently, however, there was not enough nuanced work about older adults' technology uses to make this possible as a book-length manuscript. Now that older adults are online in much greater numbers and researchers are starting to catch up with them, it is time to decipher what scholarship has uncovered—and to add to that body of research through new work designed to enhance the findings in this book.

We consider ourselves students of psychologist Howard Gardner when it comes to the methodology for this book (and other things besides). Gardner is perhaps best known for his theory of multiple intelligences, which posits that people have several types of intelligence, such as linguistic, logical-mathematical, spatial, and interpersonal; these multiple intelligences are relatively independent of each other.[24] In 1981, Gardner was among the first people to be honored as a MacArthur Fellow—commonly known as a "genius grant"—and has advised or led a wide array of MacArthur Foundation-related projects since then. We follow his thinking on the type of methodology behind this book. As Gardner states:[25]

> Synthesizers decide on a topic or problem of interest, often of general interest (like intelligence or leadership or creativity, to mention three topics that I have investigated). Synthesizers read, survey, and study broadly and deeply—they might well carry out surveys or interviews. . . . With time, the synthesis is viable enough to test on other knowledgeable or interested persons. And so, the process of critique and revision begins.
>
> Synthesis is neither science nor journalism. But ultimately, the synthesis needs to be tested publicly—and if it is accepted, it may well change how individuals think, and even how they behave. I think that is what has happened with the "theory of multiple intelligences" (MI theory) I developed—far more than with other topics that I have explored and then written about.
>
> What I do, as a social investigator—or, if you will, a "soft" social scientist—is tackle a large problem (or issue) and put together the best account or solution of that problem (or issue) that I can.

This book takes the form of an "original synthesis," to borrow Gardner's language. That means that we draw on many sources of data, some of which we have collected ourselves and some of which others have compiled and which we think is sound. This book is the result of years of work—but certainly not just by us. We borrow liberally from, and cite faithfully, many excellent scholars who study older adults, technology usage, and related themes such as health, well-being, safety, privacy, and so forth. We are in a deep form of "scholarly debt" to the many people who have worked on related issues over the past decades.

We explore all these issues with the goal of supporting successful aging. As discussed, widespread assumptions about older adults being clueless about technology are wrong, creating the need for such a myth-busting book. That said, widespread assumptions about younger adults are also wrong (e.g., they are not all universally skilled with technologies), so it is generally best not to make generational assumptions when it comes to technologies.[26]

As researchers and authors, we follow the data where they lead. We do "original synthesis" as we go along and seek to draw key themes together where they fit. When they do not fit neatly together or when they seem to contradict one another—which they do from time to time—we discuss that tension. It would not serve anyone's interests for us to pretend that there is consensus in research where there is none. Our goal is certainly not to create more myths and misconceptions about older adults and digital media.

A fun connection of our approach here to the topic of the book: it turns out that the work of a synthesizing mind is better suited to older adults than to younger researchers. As Arthur Brooks—dubbed the *Atlantic*'s happiness columnist—wrote in his 2022 book, *From Strength to Strength*, younger scholars in most fields are the ones with the breakthrough ideas, the ones who make the most unexpected and startling discoveries. With synthesis, it is different, says Brooks and others who study aging and effectiveness. It turns out that those who perform synthesis-oriented work are often older. It appears that we do accumulate wisdom over time. All those hard-won gray hairs? They mark the experience, knowledge, and skills that older people have compared to their younger, faster-moving counterparts. It appears that the kind of thinking that leads to excellent synthesis is often better done when we are older rather than when we are in the earliest part of our careers.[27]

In the chapters that follow, we introduce people who have lived experience with technologies as they age. These vignettes—mini-stories—are meant to bring the data to life. As many researchers have found, readers often remember what data mean when these are offered up through stories. Or as Richard Powers wrote in *The Overstory*: "The best arguments in the world won't change a person's mind. The only thing that can do that is a good story."[28] We adopt that strategy in this book with the goal of making the most important points memorable to you. The vignettes are based on the real experiences of people we have spoken with either in our personal networks, for the purpose of writing this book, or through studies Eszter's team conducted in various countries including Bosnia, Hungary, the Netherlands, Serbia, Switzerland, Turkey, and the United States. We have altered some people's names and some facts for privacy and confidentiality, so don't take them too literally. The goal is for these brief stories to be memorable, to bring to life the data that can otherwise be very dry and, alas, forgettable. The essential points remain—in the text and, we hope, in your memories.

On a related note, our goal is to translate complex academic literature to accessible prose with clear prescriptions. To this end, we do not present detailed statistical analyses even when we draw on statistics to answer some questions. We do want to be transparent about our methods and recognize that some people will want to know the source of our evidence. The appendix offers more details about our studies, but here are a few words about the surveys we administered to address many unknowns about older adults' digital media experiences.

We conducted a significant new study for the purpose of bringing up-to-date, relevant findings to this book about how older adults use new technologies. We also draw on findings from other helpful national surveys such as the ones conducted by AARP (formerly the American Association of Retired Persons) and the Pew Research Center.[29]

We surveyed a nationally representative sample of 4,000 adults aged sixty and over in the United States in fall 2023 through an online survey platform. We asked several questions about people's general well-being, their sociodemographic background, and of course their technology uses. These data form the basis of most figures presented in the book. The chapter on adoption also draws on another survey we administered in summer 2023 to 2,505 adults of all ages. The appendix includes information about who participated in both survey studies plus more details about data collection and analysis procedures. It is worth noting that by conducting these surveys online, we inherently screened out people who do not use digital technologies. To that point, our survey findings represent only wired older adults. Some of the lessons learned, however, also apply to those who are not online at all given that we address what circumstances encourage older adults to embrace technologies, whether they already use them or not.

The issues that older adults grapple with are not specific to one geography or culture. That said, the data on which we can draw do not cover every part of the world's population equally well. As in many areas of digital media research, the bulk of the research comes from populations in North America, Europe, Australia, and parts of Asia. These themes, of course, have resonance for older adults from all parts of the world;

we merely point to where the data tend to come from in the studies on which we rely.

We have organized the chapters that follow to start with the adoption of technology. The book then addresses the challenges of using those technologies, with chapters on support, safety and security, privacy, and misinformation. We then look at some of the possible benefits for well-being and learning. We end with specific advice about how to achieve widespread technology benefits addressed at older adults themselves, those who care for them, organizations that support them, technology developers and companies, journalists, and policymakers. We also summarize the top ten takeaways from the book on the last pages. Our principal goal is to point to the ways in which technologies can be a part of the successful aging of older adults. None of these issues or opportunities are unique to older adults, however—and, as such, we hope findings about them can help us all thrive, no matter our age.

2

Adoption

ARE OLDER PEOPLE LESS LIKELY TO USE NEW TECH?

"I keep in touch with my kids almost every day," explained Ivy, an eighty-one-year-old woman who lives in Hungary with her eighty-four-year-old husband of many decades. She is in good health, travels internationally, has family members around the globe, and uses digital technologies frequently to connect with others. Because she lived in multiple countries during earlier stages of her life, she has many friends abroad with whom she wants to keep in touch. Additionally, her grandchildren live across borders, making technology-mediated communication crucial for regular family connection. To achieve this, she uses WhatsApp and Skype to message and make video calls. She also uses older technologies like email and the phone.

Just before our interview with Ivy in 2018, a Skype update caused her and many others to lose their contacts. Frustrated by this, Ivy switched to WhatsApp, which was on the rise anyway. Given that chat and video were rolled into one experience on WhatsApp, she found it more convenient. As Ivy and others told us, WhatsApp won out through its *ease of use*. Ivy felt incentivized and empowered to try new technologies, to switch applications, and to stop using services that no longer met her needs.

What choices are older adults making about adopting certain technologies and not others, disconnecting from some that they have used, and switching from one to another? How do older adults decide what services to use and which to abandon? When do they switch, and why? A decade or so of research gives us an increasingly clear picture of how older adults make these important decisions.

One takeaway from our research, which Ivy's case also makes plain, is that older adults are clearly capable of making their own decisions about tech adoption. This counters media portrayals of older adults as somehow helpless tech users. Sure, they may need support at times, but so does everybody else.

Some of the choices users make about new services are better than others in terms of which technologies will most effectively serve their needs. One reason we chose this simple opening interview selection is to highlight the relative ease with which older adults can get started with new technologies and switch when their existing arrangements no longer work for them—and why they often make this type of choice. As the research demonstrates repeatedly, older adults often have concerns about the cost of new technologies, the ease of use, losing their contacts, or privacy. These concerns do not have to get in the way or be the end of the story, however. They simply show that older adults are thoughtful about how they incorporate digital media into their lives, a careful approach that can benefit people at any age.

Older adults need not feel stuck once they have made their initial start with a technology choice. We have found they routinely leave a particular technology or service behind entirely if it is not worth the hassle or cost. It can be helpful for those who are helping them (individuals like family members or organizations like the local public library) to know how they can encourage older adults to adopt new services and possibly abandon ones that are unhelpful, tedious, potentially harmful, or just feel too risky to be worth using. This sense of empowerment for all of us is imperative.

Digital technologies offer terrific new opportunities for communication, connection, and just about any other activity these days. Before leaping into when and how older adults get started with new technologies, it is worth noting a fact that emerges from the research: older adults almost always prefer face-to-face communication to digital communication. One thing COVID lockdowns taught us, however, is that people *of all ages* value face-to-face interactions, so it should not be too hard to appreciate this need for in-person human connection.[1]

As Ivy's case shows, older adults often start using new technologies to maintain connections with geographically distant loved ones. When it is just not possible to be in the same room as someone else, digital technologies can provide a lifeline. Older adults report that the chance to share an experience in real time with grandchildren or other younger people can lead to the adoption of a new service.[2] As we will see, older adults are also likely to leave a service if those they love also stop using it—in other words, older adults often adopt a new technology for the purpose of connecting with particular loved ones and stop when it is no longer useful for this specific purpose.[3]

Even when older adults do decide that the trade-off is worth it to start using new technologies, these services are not considered by most to be a complete substitute for earlier technologies. A great example is the telephone. The old-fashioned landline telephone, for instance, still has great value for many older adults. Research shows that most older adults seek a combination of digital and analog approaches to communicating with others and gathering news and information rather than cutting over to a "digital only" type of experience. Older adults who use newer technologies still use the telephone to connect with younger relatives, as they have for decades. There is no reason these choices should be mutually exclusive—and for many older adults, they are not.

An important motivator for technology adoption is social. A quote from a sixty-seven-year-old grandfather summarized well the role of such social influence: "If the kids are not using it, then we aren't using it either." This again highlights how important social purposes can be for adopting new services, and when those are not clear, older adults may not see a purpose for starting up with new services.

Beyond the inexorable pull of connecting with younger relatives, many older adults are open to trying new digital services if they think they can gain new skills and value from them in such a way that is greater than the costs/risks they perceive in taking on something new. Value is sometimes described in terms of "usefulness" to the older adult.[4] Put another way, as with any other consumer of a service, older adults make a trade-off in their minds: are the benefits to adopting or continuing to use a new service worth the costs, financial or otherwise?

Some types of functionalities consistently rate highly among older adults. Email, for example, remains widely popular.[5] Some older adults prefer technologies that allow them time to think and write before they are required to reply. For instance, with email, the older person can receive a message on day one, think about how they might want to reply on day two, and then ultimately reply on day three or later. The format and structure of email is slower than chat-based technologies, such as texting and in-game chat technologies, for instance, which create a sense of pressure, seeming to demand a faster, almost real-time response. Other popular applications include multiple functionalities, such as messaging and video; WhatsApp is one example we have already mentioned. Other services, like Facebook and Instagram, can offer regular updates about loved ones. While people of all ages use these technologies regularly, research shows that older adults consider these attributes as desirable.[6]

Another way to approach what may appeal to older adults is to think about the functionalities that are popular with them when using their existing devices and services. On our survey, we asked whether older adults use various accessibility features such as video captioning or inverted color display. More than two-thirds reported using at least one feature, with about a third using at least three such features. Particularly popular were the use of increased text size, magnified screen/zoom-in options, and voice-to-text typing. The more such features

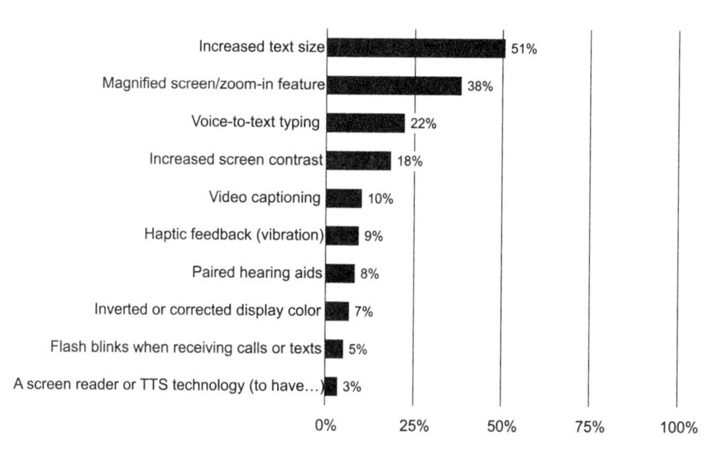

Percent of 2,000 older adults who use various accessibility features of technologies to make them easier to use (2023).

services make available to users, the more such technologies will appeal to older adults in particular.

One motivation for older adults to adopt new digital technologies, such as email and social media services, is to gain or provide support to others and to stay in touch with their networks. Reasons for adopting a new digital service include a desire to stay in touch and see pictures of friends and family, organizing get-togethers, generally catching up with others in their broader social networks, staying up to date on news and information in the wider world, and general curiosity. A simple example cited by older adults in multiple research studies is using an online application to help schedule a gathering among friends. As in other examples, this use of new technologies is not *unique* to older adults, but it does appear to be *prevalent* among older adults.[7]

When asked on our survey about whether they had ever downloaded or signed up for online services or apps for a list of reasons, half of older adults mentioned social motivations—from keeping in touch with family and friends to meeting new people such as for the purposes of dating (7% gave the latter response specifically). Other reasons were keeping up with

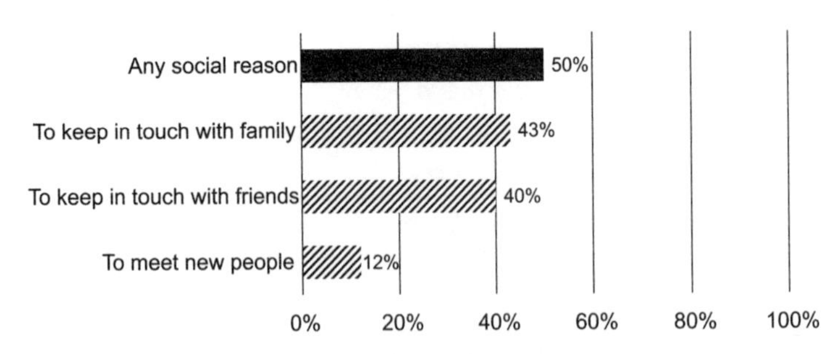

Percent of 2,000 older adults who have downloaded or signed up for online services or apps for social reasons (2023).

current events, which over a third of people identified as an impetus to start using a new service, and entertainment and fun, which just under a third picked. Fewer mentioned such reasons as making money or getting material benefits as well as the need to be more efficient with day-to-day tasks, but around 10% listed each of these reasons as well.

Some older adults perceive that using a new technological service will lead them to be less lonely,[8] although there are studies suggesting that this outcome may not result. In particular, during the COVID pandemic, older adults who used social media to connect about the pandemic reported that they felt lower levels of social connectedness.[9] This either means that those who are lonelier are more likely to flock to these services (and then their needs are not met or not met enough) or else the way in which these older adults used the services made things worse in terms of their feelings of connectedness. We will discuss the complicated relationship between social media use and loneliness in more detail in the well-being chapter.

Some older adults do not adopt new technologies because they are unaware that helpful tools exist and might be a good option for them.[10] Recall from this book's introduction that an important part of internet skills concerns the awareness of what is possible to do with technologies. If a person struggles to use one service, because, say, it lacks certain accessibility features or is too complex to understand, that person may abandon its use altogether without switching to a more user-friendly service because they are not aware that alternatives exist. Alternatively, a service may meet someone's needs in general, but if the user does not realize how they can customize it to meet their particular needs, then that may be a reason not to give it a go.

Another reason some older adults may feel barred from

new digital technology usage concerns their technological access level. Whether a household has one or multiple devices, whether these are the latest gadgets versus older ones with limited functionality, whether a user has the means to fix a device that may break, whether a home has spotty versus high-quality broadband access are all factors that may play into the extent to which older adults—or adults at any age—can take advantage of technologies.[11]

Yet others are constrained by physical or cognitive limitations. Poor eyesight, persistent pain, arthritis, and cognitive decline all appear in the literature as reasons why certain older adults do not adopt new digital technologies.[12] While some of these are harder for technologies to address with accessibility features, sight and hearing limitations are certainly ones for which technical solutions exist and should be incorporated into more services. One challenge is that even when accessibility features exist, people may not be aware of them or understand how to use them. Another is that they may not be able to afford services that are more likely to offer these features, limitations of concern to both older and younger adults alike.

Those who are aware of the tools and have the means and ability to use them may still have concerns about using them. We asked our survey participants to indicate what reasons had prevented them from signing up for services or downloading apps in the past and received some telling responses. The most common concern was the fear of their identity being stolen, which over half reported (54%). Not far behind was cost (48%), fear of activities being tracked (47%), and fear of money being stolen (40%). Three out of four of these have to do with apprehensions about privacy and security. Another privacy-related worry was fear of people in the user's social circles finding out too much information about the user, which a fifth of survey takers reported. Indeed, the scholarly literature has often identified security and privacy issues as top of mind for older adults, which is why we dedicate separate chapters to both in this book.

There were additional reasons for not downloading apps such as not knowing how to use the service or app, which just under a third of survey participants mentioned. Over a fifth reported the lack of accessibility options as a hindrance to adopting new services. These issues are a recurring theme in this chapter about how service providers can better accommodate older adult users and others with limitations.

In addition to perceived gaps in skills or knowledge of new technologies, privacy concerns, and security risks, the scholarly literature has identified concerns ranging from monetary costs to fearing the loss of human contact, the risk of feeling "addicted" to the technology, and concerns about not having anyone to ask for help or to communicate with on social media.[13] This list of worries very much motivated how we set up this book, and we dedicate individual chapters to discussing many of them.

Older adults make rational decisions about when to abandon services that they no longer wish to—or can easily—make use of. The reasons for leaving a service follow common sense. When older adults are blessed with better health, better eyesight, and so forth, they tend to keep using digital technologies. As their health, eyesight, arthritis, and other ailments grow worse, older adults are more like to abandon these same services.[14] Once again, the data point to issues of diversity among older adults: there is a great deal of variability in terms of how older adults use new technologies, and even an individual's experience can fluctuate over time.

We asked survey participants why they may have stopped

using various services and apps. Just under a third reported that a service or app did not function well, and another 22% answered that it did not offer accessibility options. Both of these point to technological limitations that it behooves companies to address.

One of the most oft-stated reasons for abandoning services (41%) was receiving too many notifications. This suggests lessons for companies and other organizations about how they set up the default of so many services. Many people—dare we say not just older adults—do not like getting constant announcements from apps and websites, so automatically subscribing people to frequent notices does not seem like a good strategy to building a sympathetic and engaged long-term user base. (Note that these are just the people who went all the way to abandoning a service due to these annoyances. The 41% does not include people who likely were bothered by frequent notifications but had not (yet?) stopped altogether using a service because of them.)

One possible reaction to the issue of notifications is that it is often possible to change them, so if people are bothered by them, they should just tweak their settings. The reality is that many people are not aware of such options—or do not know how to figure out how to change their settings. This is yet another example of how companies can do better in offering services that facilitate ease of use.

Changes in circumstance can also lead older adults to abandon a technology. If children or grandchildren move away, older adults are likelier to adopt new technologies to stay in touch with these loved ones. But if and when those same children and grandchildren move back to the neighborhood or the older relative moves closer to them, this can in turn lead to the older adult abandoning the services they had previously relied on for communication.[15] A sixty-six-year-old woman we talked with recounted using "Skype when my oldest son was in Malaysia, but well, since he is back in the Netherlands I don't use Skype anymore." An eighty-six-year-old grandfather used email while his granddaughter was abroad but stopped it completely when she returned from her travels. In this scenario, it is likely the case that the older adult used that technology only for the specific need they had at that time, not necessarily because they generally liked it or saw utility in it. Older adults like to use technologies to share everyday experiences with children and grandchildren, but when that is no longer necessary, they might drop the service altogether.[16]

Changes in life stage can be another reason for withdrawing from certain services. A seventy-two-year-old man mentioned to us that he was on LinkedIn but felt that he did not need it when he stopped working, saying it is "pretty much a dead account since my retirement." Indeed, services specifically focused on professional networking will be less relevant among older adults, many of whom have retired. While they may not technically delete their accounts, as the quote above notes, for all practical purposes, it is dead.

Negative social experiences may also propel people to abandon a service. We talk about such interactions more in the well-being chapter, but there are a few nuggets relevant to share here as well. We asked our survey respondents whether they had ever quit using a social media platform entirely in response to experiences they had on them. Over a fifth of participants said that they had indeed done this. While just over a quarter (26%) had reported using a platform less, 21% went all the way to abandoning a platform altogether. On the one hand, this is a reflection of the unfortunate experiences people can have on social media. On the other hand, it shows the agency that some older adults take in abandoning services that they do not believe to be good for them.

The alternative to abandoning a type of service is to switch to another one that offers similar functionalities, but in ways that meet the user's needs better. Indeed, the most popular response at 44% from survey participants about why they have stopped using services was that they found another one.

Ivy's experience at the start of this chapter showed why she and her social circles switched from Skype to WhatsApp. Skype had rolled out a new version that resulted in many users losing their contacts, making it hard to connect with loved ones. This experience of switching from Skype to other communication services came up in several other interviews across different countries. One sixty-five-year-old woman explained she used to use "Skype in the past, but somehow, as I said, because now those who live abroad have switched to Messenger most frequently, I go with the flow." Another older adult man noted: "We have completely gotten rid of Skype lately, it has become completely unnecessary. In general, people who we are in touch with use WhatsApp, so we don't use Skype with them. We only use Skype with those who don't have WhatsApp."

WhatsApp offered easier access and made both messaging and video chat conveniently available, so it was a helpful alternative to many. The fact that older adults will often switch to other services rather than abandon a certain type of online activity altogether shows how they make decisions about preferred services.

Technologies could do more to make switching easier. There is an obvious reason why technology companies might not want to make this easy, of course. After all, technology companies make money when people use their services, and they do not want to encourage people to leave by making it simple to do so. For this reason, it is highly unlikely that for-profit companies are going to be so enlightened as to make switching away from their service to another service easy for older adults or anyone else.

There is a clear role for government to play in nudging companies to give users access to their own data. Recall Ivy's case at the start of this chapter. She and her friends ended up switching from Skype to WhatsApp since they had lost all their connections during a software update. This was sufficiently frustrating for them to abandon the use of the service. Access to one's data should be in formats interpretable by the general user. Currently, even on the rare occasion when sites give users the option to download their content, the output is usually in a format that most people cannot utilize. This relates to the general idea that older adults need to feel autonomous when making choices about their technology uses. While many have the right attitude, the technology can stand in the way of their ability to pursue their preferred uses.

The research about older adults and their use of technology underscores a commonsense truth: when systems are simple rather than complex, they are easier for older adults to use. As in many other elements of this book, what is true for older adults is true for all of us. The simpler the service is to use, the more likely it is to gain widespread adoption and longer-term commitment.

This basic insight about simplicity over complexity should guide government policy with respect to technology regulation that favors older adults. We propose that governments require technology companies to make it easy for older adults to control their personal data as this will add to their sense of autonomy when it comes to technology use. The mechanism for deleting their personal data needs to be extremely clear and simple in order to fulfill its function. Governments can help with certain types of mandates to support older adults in their technology usage.

For instance, governments could pronounce a policy goal to make it easy for people to switch services from one to the next. Lawmakers could consider requiring that technology services design themselves to be interoperable with others.[17] To the extent that systems can interoperate, they ensure that data can flow from one to the next with ease on the part of the technological system's user—in our case, an older adult. There are many ways to design a system to be interoperable, and a mandate would have to be drafted with care to achieve its goal without too many costs and unexpected consequences. But such a set of rules could help older adults enormously.

Here is how it might work. If an older adult were to join a social network service that requires them to enter their real name, their birth date, and various other facts about themselves, the rule might require the technology company to collect and store these data in a way that would make it easy for the user to move the data to another service. Imagine that the older adult includes some pictures and some text in their profile on the social network service, adds some contacts among their friends, and so forth. Next, imagine that the older adult wishes to leave that service for one reason or another and wants to move these data off the first social media service's platform to another. Again, recall Ivy's experience: she and her friends were confused by the loss of their contacts, something other people in our interview study also mentioned as an issue. Before enforcing a major update, platforms should advise their users to export their contact details so that they have these on hand in case the update renders these unusable.

If a government mandated that the social media service had to promise two things—that users could easily delete anything they shared on the service and could move the data to a new service with the click of a button—older adults might be more likely to join up in the first place. If it would be a simple matter to pull all their data from one social media service and load up the relevant data to a new one when the time came to switch, then older adults might be more likely to use the service that best served them at any given time. Worth noting, as with so many other provisions, such a government mandate would serve the interests of not only older adults but users of all ages.

A government mandate in favor of interoperable data systems for social media services or other online systems could be accomplished if companies agreed to a common standard for how personal data are collected, stored, and handled. The use of this standard could be managed by a third-party nonprofit or a government agency. No matter what form the rule or the standard might take, the mechanism for switching between services must be especially simple for it to work at scale.

Social media are some of the most popular online services, and much of this book deals with their uses, so it is helpful to establish how popular they are among older adults. On the whole, the vast majority (90%) of the 4,000 older adult survey respondents reported using at least one such platform, and two-thirds used two or more.

In another survey study, we asked 2,505 adults of all ages which social media they use, allowing us to compare older adults' experiences with those of adults under sixty. (The appendix describes this study in more detail.) Among older adults, Facebook and Pinterest are the most popular platforms. The portion of people who use these two platforms is not any different by age; they are similarly popular among people under sixty. On every other platform, adults under sixty are more likely to be present, with especially large differences in the popularity of Instagram and TikTok.

When it comes to active engagement (posting, sharing, or commenting) on these platforms, older adults behave in

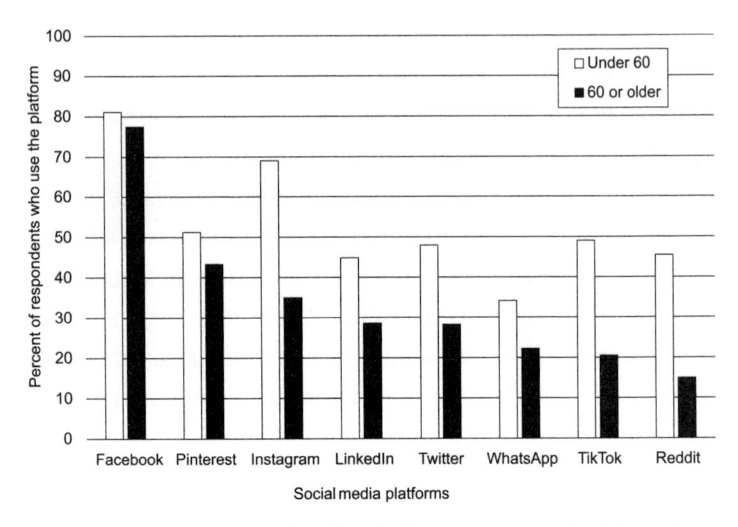

Popularity of various social media platforms among 1,698 adults under sixty compared to 807 adults sixty and over (2023).

remarkably similar ways to those under sixty in most cases. Instagram, TikTok, and Pinterest are the only platforms where those under sixty are more likely to engage actively than older adults (among those who use each respective platform). Statistically speaking, older adults are not any less likely to post, share, or comment than their younger counterparts on Facebook, WhatsApp, Twitter, LinkedIn, and Reddit.

A major theme we keep returning to in this book is the variation in digital experiences across different older adults. While the popular myth may be that older adults are all essentially hapless when it comes to the use of new technologies, the truth is much more interesting and complex. There is a lot of variation within the population of older adults in terms of the adoption of new technologies. For instance, when it comes to social media, those who are in their eighties and older are less likely to adopt a service such as Instagram as compared to those in their sixties and seventies. Interestingly, for Face-

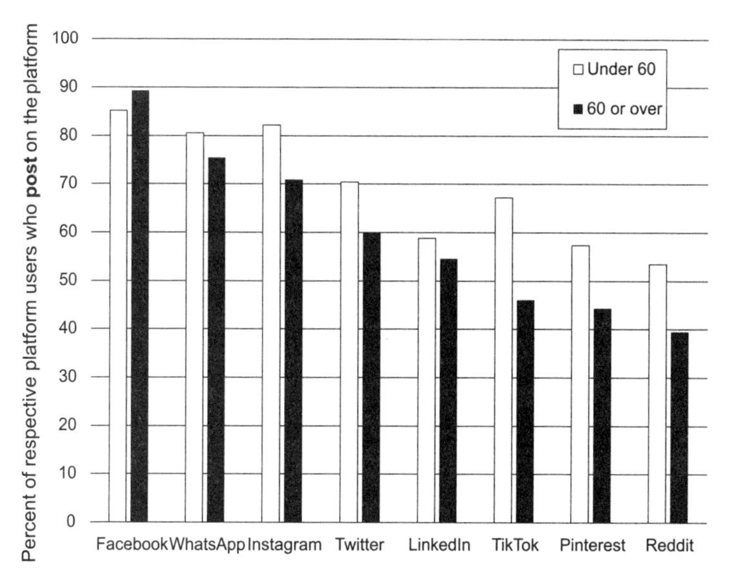

Proportion of adults under sixty compared to adults sixty and over who post on various social media platforms, among users of the respective platforms (2023).

book, there is little such variation even when considering the oldest old. One explanation for this is that older adults are more likely to adopt an online service that is more established and text-based rather than newer and image-based.[18] This is clearly reflected in our analysis of Instagram and TikTok above, which show the largest variations by age group.

As noted earlier, with changes in employment status come changes in the use of platforms that are focused on professional contacts. Those who are still in the workforce have more reason to use a site like LinkedIn than those who are retired.[19] Indeed, among our survey takers, those employed were over twice as likely to be on that particular platform compared to retired respondents.

In the first chapter we mention the concept of the "digital divide" or the haves and have-nots of technology use. Such a divide exists even just among older adults.[20] There is a fair amount of diversity in terms of who uses new technologies and who does not among over-sixties. Trends in adoption of new technologies can vary by geography and even by study, but it does seem that adoption of social network services, for instance, is correlated with factors such as education, wealth, and health—much as it is within other age groups.[21] In a study of older adults in Chile, for instance, the most important determinant of whether an older adult used digital technologies was their education level.[22]

Those with better internet skills are also more likely to adopt social media.[23] In our survey of Americans sixty and over, the vast majority of the most skilled users (93%) were on a social media platform compared to 83% among the least skilled users. While both 93% and 83% are high numbers, there are clearly differences between the two groups. These differences are especially pronounced with some specific platforms such as Instagram. Among the most skilled, 48% are on that platform compared to just 23% of the least skilled older adults we surveyed. And to clarify, these skills are unrelated to social media uses but represent more general internet skills.

Variations in skills give some clue as to why education is related to different rates of social media adoption. Those with less education are less skilled with the internet. Those with less education are less likely to be on social media, as are those with lower internet skills. When we simultaneously consider how people's education and skills relate to their social media adoption, it turns out the real driving force between differential social media adoption is general internet skill not education. Those who understand the internet better—unrelated to social media—are more likely to start using social network services.

The current reality is that some older adults find it easy to sign up for a doctor's appointment online, get a free vaccine at a drugstore by signing up on an application, or get free health tests sent to them through a government website while their neighbors can do none of these things. It should not be the case that some people can take enriching online courses, use YouTube videos to do yoga on a floor mat at home, and even teach others via Zoom while their neighbors are kept out of all these activities based on their education or skill level.

In a perfect world, everyone would be able to jump on a new service or website to get done what they need to get done. That is simply not the world we live in. There is a great deal of diversity within older adults in terms of skills and abilities to use new technologies. These differences mean that the gap between those who are best positioned to thrive in a digitally mediated age and those who are not will continue to grow unless targeted steps are made to address them.

A number of things might be done to drive and sustain adoption of new technologies by older adults. A virtuous circle can work wonders. When family and friends recommend a technology and help each other get started with the technology, use the technology together, and continuously troubleshoot issues, older adults are likely to stick with a technology.[24] Support systems can help drive a sense of self-efficacy on the part of older adults[25] and encourage adoption.[26] These same support systems can ensure that older adults get the positive benefits of new technologies while mitigating potential harms.[27] We explore this in more depth in the next chapter.

To ensure that older adults do not abandon a particular tool or service, the answer is likewise fairly straightforward: training and support can make it less likely for users to abandon technologies.[28] A counterexample from the research is instructive; complex wearable devices that might be good for health and safety may not get much use if they are perceived to be too complicated or frustrating to use.[29] While training and support can help, companies need to do better in the design of the systems themselves.

A recurring theme in this chapter has been the importance of accessibility options. Service providers must make their systems intuitive and simple enough to ensure that a diverse group of older adults can feel confident in their ability to use the technology. What is the incentive? The average older adult has more disposable income than the average younger adult, so this is a demographic that has resources to spend on technologies, assuming those technologies are sufficiently user friendly for older adults to add them to their daily routines.

Too often, people who are in a position to make good suggestions to older adults that could help them use digital technologies more effectively hold back because they mistakenly think of older adults as averse to new tools. This is a mistake. Research shows that when older adults understand how a new service can be helpful to their lives—for example, by connecting them to loved ones—they are often very open to trying it out and making it part of their everyday lives. Much adoption happens thanks to people in one's networks using new technologies, so talking about them and explaining the incentive, such as access to more contacts or more content, can be a helpful gesture.

Technology companies have a key role to play as well. The greater the ease of use and sense of control, the more likely that older adults are going to start using and continue to use helpful technologies. If older adults actually enjoy the experience of using the technology and perceive it to be useful, they are often willing to put in the work to adopt and continue using it.[30] But if their fears about privacy concerns are reinforced, or they struggle to make the technology work without outside help, they can give up. When older adults want to switch services, service providers should make this easy for them to do by giving them easy access to their user data. If companies will not do this on their own, then government policy should step in to make such interoperability accessible. Those who design and offer technologies to older adults would do well to heed the exceedingly simple lessons from the research about older adults and their patterns of technology adoption, use, and abandonment: make it accessible and transparent as to the benefits of its use.

As we hope to do in this book, some older adults are fighting ageism and stereotypes. Numerous popular TikTok accounts feature older adults engaging in activities that often make younger users popular, such as dancing and humor. The @grandpachan account is of a couple whose description states: "We're 82yo Korean grandparents making things to connect to our 4 grandkids." The "things" here mainly refer to lighthearted short dance videos. They have garnered over 30 million likes from hundreds of millions of views. Also through music and dance, seventy-four-year-old veteran twins (@hanelinetwins) from North Carolina are fulfilling a lifelong dream of entertaining others with over 2 million followers and 36 million likes. The @nannabea account has over 50 million likes and shares the experiences of a grandmother of four in Britain. The @oldgays account handle speaks for itself; it has over 10 million followers with over 225 million likes. It features the lives of four friends, aged sixty-eight to eighty, living in small-town California.[31]

This is the type of social media use that care facilities can also embrace to get their members connected with the wider world. Arcadia Senior Living Bowling Green in Kentucky is such a community where staff have worked with residents to create fun videos, some of which have gone viral. Their most popular clip features several older women recreating a segment of Rihanna's 2023 Super Bowl halftime show. This clip has garnered over 40 million views with over 4 million likes. The care facility's account (@arcadiasrlivingbg) features numerous residents often working together on a video, suggesting engaged behind-the-scenes preparations.

Some TikTok accounts from older users are specifically about life during one's mature years,[32] while others are about general humor, dancing (sometimes in a multigenerational arrangement), cooking, and a myriad of other types of content. Research on why some older adults have embraced TikTok suggests similar motivations to other media uses: they help foster social connections.[33] Our point in mentioning these examples is to highlight that some older adults are just as active—if not more!—as younger ones on numerous platforms. They embrace technologies in various parts of their lives and use them in informed ways. There is nothing inherent in an older adult being averse to technologies, and the benefits they can gain from welcoming them can be immense. Some may need support, however, and it is up to their communities to provide it.

The most important takeaway regarding the research on older adults and how they get started with new technologies is the importance of a sense of self-efficacy and control. When technologies are user friendly, when they offer accessibility features that address decline in health such as limited sight and dexterity, when they are transparent about and facilitate safe and secure uses, then they encourage uptake by older adults (and presumably everyone else). In all this, support can make a huge difference, and we take that up in the next chapter.

3

Support

HOW DO OVER-SIXTIES SEEK HELP?

Norman and Dale had a fairy-tale romance as college students in the 1950s. They went to football games and plays, concerts and restaurants. They had a storybook marriage too—they were together for more than sixty-five years. They each had good, steady jobs working at universities; they loved one another; and they adored their apartment in New York City's West Village. They had many friends, from across New York and all over. They never had children.

In their late eighties, their health began to fail them. Sadly, Dale passed away peacefully just before they reached ninety.

Norman, for the first time in more than half a century, found himself living on his own. Still in his Greenwich Village apartment but now living alone, Norman's need for activities and companionship took on new shapes.

There was an additional big problem: Norman became a widower in the middle of a pandemic. COVID-19 had gripped New York City as it had virtually every place in the United States and much of the world. It was essentially impossible for friends and family, nieces and nephews and godchildren to visit him. Caregivers could visit the apartment if he had a medical need; and New York's ample set of delivery services brought him any kind of food he could imagine. But interpersonal social connection was effectively impossible to come by in Norman's greatest hour of need.

Norman had been a professor of medicine his whole career. He knew his way around technologies related to his field and research. He had long used a computer and had a simple cell phone in his profession. But the digital media that made new forms of interconnection and access to knowledge possible during a pandemic? That was quite another matter.

Over the phone, Norman could reach his loved ones as he typically had in the past. Those with a few minutes could walk him through the basics of placing an order on DoorDash or Grubhub. He mastered that skill well enough to keep himself fed even as pain in his knees and spine made it hard for him to get outside for a walk or a COVID-safe takeaway drink at his favorite café. (Norman still preferred to tip using the cash he kept in a drawer; once a waiter and busboy in what he called the "Borsht Belt" of upstate New York during his college days, Norman insisted that the delivery persons would prefer cash.)

For entertainment, Norman loved mystery shows: *Agatha Christie's Poirot, Inspector Morse* and *Inspector Lewis,* P. D. James's *Dalgliesh,* almost any of the British ones he could watch all day. The advent of Netflix and Amazon Prime made that entirely possible. With a little help, he was able to hook up a television set to stream his favorite shows as much as he liked. He needed to have the volume up as his hearing was going, but that was no problem since he now lived alone.

Beyond ordering food and watching shows, Norman was at a bit of a loss when it came to digital media. Used to the phone, Norman had not gotten into other forms of social interaction using new media. Family gatherings over Zoom were not going to work for him as that was unfamiliar terrain. He did not have a Facebook profile or use Instagram or Twitter.

Their whole lives, Norman and Dale had made regular trips to New York's wonderful public libraries. Every year, they had dutifully contributed to the Friends of the Library funds. On his own and no longer able to make the several-block trip (even if the library had been open during COVID-19), the wonders of what a library could offer him using new media seemed too hard to access.

The endless possibilities of connection and stimulation of the digital age that so many people benefit from and even rely on were just out of reach for Norman without a little more support. Without using more services and knowing how to find people on them, he could not rekindle long-lost friendships. Without appreciating the myriad of online communities that flourish around every hobby and interest imaginable, he could not make new connections with like-minded people. While he could manage with the basics, he was missing out on what might have been possible in his late eighties, newly on his own.

It is a great blessing to have older adults in one's life. As loved ones—parents, grandparents, uncles, aunts, mentors, friends, next-door neighbors—grow older, caretaking roles change. Children who were the ones cared *for* take on new responsibilities as they move into adulthood and those who supported them are the ones now in need of support. Providing guidance, tips, and encouragement to older adults as they access new forms of digital media can be some of the most useful ways such support can take shape, whether coming from other older adults or younger people.

Anyone who has ever tried to help with technology use

also knows that it can be some of the most frustrating topics on which to interact with older loved ones. Digital media can seem baffling sometimes. Printers not connecting. Passwords not matching. Programs not loading. Edits not sticking. Banking not working. Volume not changing. The benefits for interaction and access to ideas and people can seem tantalizing, but annoyingly just out of reach. That in turn can result in frustration when well-meaning assistance falls short of the goal.

The typical media portrayal of older adults and their use of technology paints them in a negative light or tends to be patronizing. Think of the headlines calling out this group for being behind the curve or clueless. These images are neither accurate nor fair. Nor is the image that older adults form a monolithic block, all struggling with media in the same ways. Anecdotes and cheap shots add up to a portrayal of older adults using digital media (or not using it, for that matter) that obscures what is really going on. With such stereotypes, is it any wonder that some older adults themselves believe that they cannot master new technologies?

Yet research has shown that some older adults are more skilled, more able to learn and adapt, and more curious than we give them credit for. There is a myth that young people are "born digital" and that older people cannot use technology.[1] It turns out that neither part of that statement is true. There is large variation in internet skills among younger people with some as savvy as public rhetoric assumes, but many lacking in important domains such as identifying the sources of online information or creating privacy settings that match their preferences.[2] At the other end of the age spectrum, some older adults are very knowledgeable and comfortable with technology and can serve as support for their peers.[3] Perpetuating age-related myths about technology use leaves both groups shortchanged.

There is a growing body of research that provides a road map for how we can best provide support for older adults in our lives as they interact with digital media. There is more good news: older adults are often eager to learn, but just need a bit of guidance.

As we note in the previous chapter, older adults are a highly diverse group when it comes to their use of new media. Many older adults use new technologies regularly and do so effectively. Of course, there are those who do not use new media at all and are scared to try. Research also shows that there is a wide degree of different support needs and different ways we can offer support.

Diversity abounds: just as older adults can vary greatly in their support needs, the types of support that may be required and that are possible to provide can take many forms. Support can cover hardware and software—right from the point of opening the box or downloading a new application. Support may be needed to navigate websites, fill out an online banking application, submit a government form, or operate Zoom. It could mean helping with a library e-book service or purchasing a subscription to Hulu or Netflix. Support might also take the form of configuring, charging, or replacing a mobile device. Such basics as changing the volume or turning off sound altogether can befuddle some. And who has not struggled with a printer or other peripheral? The skills needed to provide all these forms of support are unlikely to reside with any one support provider.

The challenge is to understand what needs older adults have and whether those needs are being met. If the answer is no, then we have to figure out the best way to make that support available as older adults are increasingly incorporating digital media into their lives. On the flip side, as many older adults are retired and have more leisure time, they may find online opportunities that others of us are not privy to and can learn from. The support can go both ways.

Do older adults need help with technology? Does that seem like a silly question? Popular belief holds that they all certainly must, given that they are clueless about it. The reality is more complicated. When we asked survey participants whether in the past year they have had questions about or run into problems with using technology, a third told us no. This resonates with some of the reactions we have gotten to the book from friends who think we are writing about the tired trope of the naïve older user. We then clarify that an important goal of the book is to explain that older adults are much more diverse than that when it comes to their technology experiences.

We asked survey respondents whether they have had questions about or run into problems with technology such as downloading apps, using video services, learning new programs, and sharing content in the past year. Just over a third said that they have had questions or problems with technology "once or twice," with 16% claiming "three or four times" and the remaining 14% "more often." There is some variation in this by age: those in their sixties were more likely to say that they never needed help while those in their eighties were more likely to say they needed it more often. But the differences are not huge. There is no obvious difference either by education or financial hardship. The one clear link is that those with lower skills are more likely to have run into questions and problems.

When it comes to digital media, the support needs of older adults often—though not always—start right at the beginning. As soon as an older adult gains access to a new device or service, they may need a hand or a word or two of advice or encouragement. About three-quarters of older adults report needing support to set up their devices as soon as they receive them.[4]

Much can turn on whether and how older adults receive support in setting up their new technological devices and getting going. As with people of all ages, if you get help right away and learn how to go online, you may be more likely to spend more time doing so in the future. Those who do not have support off the bat may experience this lack of support as a hindrance to getting comfortable and confident with going online, discouraging longer-term use.

This insight about how things get started has to do with a theme that recurs through the research: the concept of confidence. When users get the kind of support with digital media that builds confidence, it can make a big difference. Confidence in access to help with technology is positively related to internet use among older adults.[5]

When older adults are confident about either their own skills with digital media or their access to help, they tend to gain more from their experiences and build from there. Where older adults get frustrated quickly, lacking confidence in themselves or in those who might support them, the opposite effect can kick in. As the old phrase goes, the end depends on the beginning: the first point of contact can make a big difference. Providing the type of support that builds confidence is particularly significant.

After establishing how many people have needed help with questions they have had about technology, we asked survey participants to indicate how they have tried to get such help. Although doing an online search and experimenting on their own are the most common approaches, surprisingly many older adults do not turn to such strategies. In fact, a

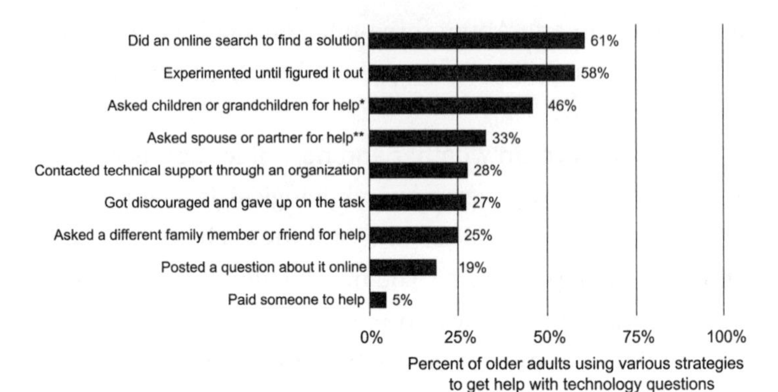

Prevalence of ways 2,000 older adults have tried to get help with their questions about technology (2023). (*Among people who have children or grandchildren. **Among people who have spouses or partners.)

full quarter of older adults reported doing *neither* of these when they need help. Asking close family members (children, grandchildren, spouses, partners) is the most popular form of social support requests. At under 20%, comparatively few people have turned to an online community to find answers, and very few (5%) have paid someone for assistance.

Relatively rare was asking a more distant family member or friend for help, with only a quarter of respondents saying they did this. There is a mismatch here in what support may be available but is not tapped, again likely due to stereotypes. Challenging the widely held belief that older adults are not good with technology, we encountered active and savvy users throughout our research. In particular, we heard an openness to offering help, but some did not experience such requests. An eighty-four-year-old man in the Netherlands summed up his related experience of being available to offer help: "People don't ask for it so I haven't done it. If they would ask, if I could help, I would." We had the following exchange with a seventy-two-year-old woman in Switzerland when asking whether she ever helped anyone other than her husband:

INTERVIEWER: Do you ever help anyone else, other than your husband, with doing things online?

PARTICIPANT: No.

INTERVIEWER: Do you think you would help if someone asked you for help?

PARTICIPANT: Sure.

INTERVIEWER: You would feel . . . comfortable.

PARTICIPANT: Sure. If they needed my not-so-advanced level of help, yeah I would do it. Yeah, I would try.

Stereotyping older adults as helpless with technologies strips them of opportunities to support others with their needs.

We also asked on our survey how often older adults have been able to find a solution to their questions about and problems with technology. Less than a fifth said this happened all of the time, and another 43% reported it happening more than

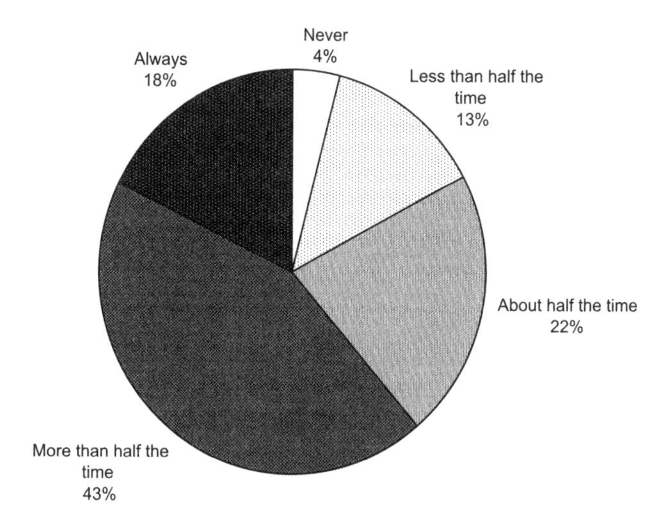

How often a sample of 2,000 older adults have been able to find a solution to their questions about and problems with technology (2023).

half of the time. The remaining 39% was left without an answer at least half of the time. It is also notable that 27% said that giving up is one of their responses when they need assistance. These figures suggest considerable need for improved support.

Research gives us some important pointers about the importance of *how* the support is given. It is clear that support givers' approach to older adults can affect the extent to which those in need keep asking and keep trying to learn themselves. It is crucial to bear in mind that the messages sent by those providing support can have long-term consequences for those they are trying to assist—especially if older adults get turned off in the process.

Some older adults experience frustration while they are receiving help. If they ask someone for assistance and the interaction goes poorly, they may not want to ask the same person for help another time. This was the case with an eighty-five-year-old man we interviewed in Hungary. Although his grandson works in IT, the interviewee had stopped turning to him for help because he found his attitude "cumbersome." He now turns to his children (older adults in their own right) and occasionally contacts a distant cousin in Germany when he runs out of other options.

It only makes sense: if an older adult feels that they are being treated poorly or are being infantilized or derided in the process, they are going to be less likely to ask for the support they need moving forward. While some older adults report relying on younger people such as children and grandchildren for help, others report that they prefer to ask those closer in age for support rather than those who are younger, often based on prior negative experiences with the latter group.

One specific challenge the eighty-five-year-old man described was getting rid of the red squiggly lines under his text when writing in a different language. He was in the midst of translating a document from Hungarian to English in Microsoft Word. Suddenly every word was underlined, suggesting typos, since the setting was checking Hungarian spelling even though the document was now in English. (One would hope that with advances in artificial intelligence, future versions of Microsoft Word will at minimum ask the user whether a document is in a different language when the program notices that everything is suddenly "misspelled," but we digress.) When asked what he would do to solve the issue, he said "nothing." Although he knew that the problem stemmed from a mismatched language setting, he did not know how to change it.

What would you have done when faced with such a question? For many, the obvious response is to search for an answer online. With current search engine technology, something as informal as *how to get rid of red squiggly in word* will yield useful information including user-friendly video tutorials. So why did this older adult not search for an answer? We point yet again to the importance of awareness. Here the awareness concerns the recognition that most questions that may occur to someone have already occurred to others as well and thus likely have answers somewhere in some form online. But just in case the search did not yield useful results—although, given a well-targeted query, more often than not it does—especially savvy users would know to turn to relevant online communities for input.

As our survey results have shown, over a quarter of older adults have gotten discouraged and given up on a task, something that was also reflected in interviews. That is a considerable number of people not getting their needs addressed. What inspires some of us to pursue a solution with a search query while others give up in defeat? One likely motivator is confidence. Searching for answers online is a much more appealing approach for those who have the confidence in knowing how to implement the advice they receive than those who are not sure they would know what to do with it.

The idea of confidence is once again key: if an older person feels assured that help will be provided in a way that works for them, and this process does not make them feel bad or like a burden, they are more likely to use the technology in question in the future. The message to those who offer support is clear: the "how" matters as much as the "what" when it comes to offering and giving support. Kindness and patience are key. A quote often attributed to Maya Angelou:[6] "I've learned that people will forget what you said, people will forget what you did, but people will never forget how you made them feel." The idea resonates for how we think about offering support for older adults using digital media.

One more point here about awareness. After hearing that the older adult above was working on a translation, we asked whether he knew that there were online services that could do a first pass for him for free. He did not. To this end, it can be helpful to converse with older adults about their daily activities in ways that may give hints about what types of online services could be useful for them and then share those so that they can benefit.

Contrary to the view that older adults rely principally on younger people for support, research points in the other direction: toward the significance and variety of technology assistance exchanged among older adults.[7] Peers are often listed as the preferred source for technology help in this age group.[8] In other words, older adults often prefer to gain support from their own cohort, whether a spouse, friend, or acquaintance.

Older adults who have living partners or spouses are often in an advantaged position to access and make the most of digital media. The support that older adults report receiving often comes from a partner or spouse who has used a certain device or application before. Not every older adult loves having to ask their partner or spouse for help, but the immediate access to assistance can make a big difference.

The support in using digital media that members of older couples give to one another is almost certainly greater than we have understood in the past.[9] Older adults tell researchers that spouses are not a fallback, rather, they are often the first line of informal support while using digital media, followed by their children, grandchildren, and friends. Some of this is due to convenience. One seventy-five-year-old man in Switzerland who often turns to both his wife and daughter for assistance was very explicit about this: "whoever is there," referring to his home.[10] When asked what he does when no one is "there," he said he would "turn it off and wait" until his wife came home. A seventy-three-year-old woman in the United States echoed the sentiment: "I mean, it would be [my husband], only because he's right here." Similarly, a sixty-seven-year-old man in Switzerland noted: "If my wife is close by, I can ask her."

Sometimes one spouse is more skilled than the other, which motivates the partner to approach them for help. A seventy-one-year-old man in the Netherlands referred to the relative expertise of his spouse: "My wife had worked with the internet before so she is the wise one of us two. If I have a problem doing something, she can solve it, nine out of ten times." When asked if anyone helps him use the internet, a seventy-seven-year-old man in Hungary replied, "Yes, my wife. She started using it before I did. Actually, in every way concerning the internet, she was always ahead of me." A seventy-three-year-old man in Switzerland turned to his wife even when she was not at home: "If I do have a problem, usually my wife is a little better than I am, and so I call her." While gender stereotypes about technology skills might suggest that men would be the ones providing more support, studies have not observed such a dynamic among older adults.

Taken to an extreme, some older adults *only* access digital technologies through others, a practice referred to as proxy use.[11] We encountered an example of such proxy use when we spoke with a sixty-three-year-old man in the United States who was still using a flip phone in 2023. He shared that although he accessed email on a computer, he did not have social media accounts because with Facebook "you open your front door to the public." When asked whether he uses the internet beyond email, he explained that his wife looks things up for him. There are downsides to proxy uses. One is that the person relying on others rarely picks up skills themselves. As a result, if the support giver is no longer available, the older adult is then completely cut off from those resources.

Norman and Dale's story that opens this chapter also points to the importance of focusing on the social context for older adults when thinking about their technology support needs. For older adults who have other older adults with technological confidence and skills close by, the support needs may be quite different from those who do not. And when a life change occurs, a change in support may be needed sooner rather than later.

The support that some older adults need is actually mostly to do with getting themselves up to speed and feeling confident that they can help themselves or find help for the trickiest problems. For some older adults, they need the confidence to get going, but they then want principally to do it on their own. The key aspect of support is at the front end of the process: getting going, building confidence, and identifying places to go for future support. While, for many, doing an online search or playing around with a device is often the first strategy to finding a solution, our survey results showed that about 40% of older adults will not do one or the other, and a full quarter will take neither strategy when faced with a problem. A goal then is to build older adults' confidence so that they believe that they are equipped to find solutions themselves.

Digital tools have a quality that some, although clearly not all, older adults recognize: you can try something over and over without consequence in most cases. One study investigating how older adults learned to use their tablets specifically mentioned "playing around" as an important way users could acquire digital skills.[12] In many aspects of the digital world, there is little to no cost to trying and failing, unlike with certain things in the analog world. Consider a home improvement project. If someone puts numerous holes in the wall (not to mention dismantling a wall) that do not end up meeting the needs of the project, there are visible consequences to the mistaken approach that may require considerable effort and cost to fix. But with digital tools, a mistake often does not "consume" anything. If there is enough patience and confidence that eventually things will work out, an older adult can try and try again until they have figured out how to manage the task. Communicating this aspect of digital technologies to older adults is important so that they feel comfortable experimenting. (There are exceptions, of course, like when a certain action could result in security violations. We address these in the chapter on security.)

Several older adults we interviewed were happy to figure things out by themselves.[13] When asked who helped him when he first started using the internet, a sixty-eight-year-old man in Hungary simply stated, "Well, I did." In Switzerland, several people we talked to said they "learned by doing." When we asked a seventy-three-year-old man in Hungary whether he needed help using Facebook, he replied, "No, it can be figured out if you have been using the internet for a long time," which he had done for work before his retirement. An eighty-four-year-old man from the Netherlands shared that he was glad to have help learning the basics, but the rest he wanted to "discover" for himself. "If I do it wrong, well then I do it wrong, then I will start again," he said. "It is a device where you can start over and over again."

Many others we interviewed described similar sentiments. They reported reaching out to younger relatives or more savvy peers of similar age to get help, while also desiring to understand the technology more deeply themselves. The mix of knowing how to get help and having a way to do so, combined with a desire and ability to help oneself, offers a vision of what can be possible for older adults living in an increasingly digitally interconnected world.

Most commonly, older adults report that they use a mix of support to meet their needs. They might start out asking their spouse and end up paying for technical support at a store, going to a help desk where they bought the device, or seeking assistance at the local library.

Not everyone is in a position to pay for support with their technology needs, and very few older adults do so. For some older adults, though, where money is available, paid help can provide an essential supplement to the informal types of support that are easily available from family and friends. Some reported not wanting to bother their younger relatives; others got turned off by the initial reactions they received when they asked for help at previous times. Yet others reported that they knew enough to ask for specialized help for highly technical needs.

The more expertise an older adult has, the more they may

realize that they need formal assistance in some realms. A sixty-six-year-old man from the Netherlands who still works in the IT field reflected on how he deals with security issues that he cannot address himself: "It's not a matter of fiddling a little . . . I use it professionally. So it's my job. And if that thing breaks down, I can't work and I can't earn any money."[14] Accordingly, even though this man is an expert with many aspects of digital technologies, when it comes to security, he relies on professionals for assistance. Computer security stands out as an area where formal assistance may be especially useful for older adults, and we explore this specific case in more detail in the next chapter.

An essential skill for older adults, as for all technology users, is to know one's limits and recognize when external help is the best route to a solution. The ability to pay for help is a luxury that some older adults can afford, but others cannot. Support for computer security, for instance, is something that is very hard for even an IT professional who is working in another part of the field to master as per the example above. Support via a paid service, a library, or a community center is likely to be a better answer than for an older adult to try to figure out the emerging best practices of cybersecurity on their own.[15] Since people of all ages have related needs when it comes to cybersecurity, for communities to invest in this type of widely available support seems like a worthwhile strategy whose benefits will reverberate well beyond older adults.

Case in point, many public libraries have embraced the digital turn not just in terms of the cultural products they offer—e-books being an important example—but also concerning the support services they make available. Recognizing that older adults may especially value patience in such a context, some libraries provide dedicated time slots for this demographic. There are several models for providing technology support. One form is to offer workshops or classes where a few people learn together. Some of these can be restricted to older adults only so that they can be in a comfortable space with peers. Such a communal approach can be advantageous as shy people can learn by listening to other people's questions. Additionally, sometimes new users do not even know what questions to ask, so hearing other people's queries can be a great introduction to useful information.

Another option is for a library or community center to post times when a helper is on hand for one-on-one sessions. Easily accessible drop-in sessions can be helpful for cases where the question is too small for people to organize a visit around, but they would like to have access to support when they are already on location for other reasons, such as a hobby meeting at the senior center. A third possibility is to allow for people to sign up ahead of time for one-on-one consultation (either in person or on Zoom). These are all models that currently exist at various public libraries and community centers, and they can go a long way in addressing a community's technology needs without burdening the user financially.

A few decades into our widespread use of digital media (and research about what works and what does not), a few things have become abundantly clear. The reach of technology is expanding into nearly every aspect of our lives: we use digital technologies to buy things, to connect and reconnect with one another, to get ideas for dinner, to learn languages, to access government services, to find contractors, and on and on. During the COVID-19 pandemic, that list grew to include millions of people who had to go to school over the internet and millions who gathered to pray or go to memorial services together over YouTube and

Facebook. Virtually every aspect of social and economic lives requires a connection through technology at some stage.

A key learning from the implementation of programs to increase access to technology is that you have to "be there for the last mile" to make it accessible to everyone. This expression comes from the physical building out of technical broadband infrastructure all the way to the last mile of reaching people's homes. But with the internet, even that is not enough: it is not enough to build it and then assume that everyone will be able to come. It is certainly a necessary condition to getting everyone online, but in and of itself it is not sufficient. Considerable research has now documented across countries and over time that skills and support are also a necessary component of getting everyone truly online. While it is true that many social media platforms have scaled extremely quickly—consider the reach of Facebook from zero users in 2000 (when it did not yet exist) to over 3 billion by 2024[16]—no technology is universally accessible on its own.

Those who design technological systems often focus on the essential engineering tasks. And of course, these have to be done right: a technological system is no good to anyone if it is unreliable and breaks down all the time. But typically we have not focused enough on "the last mile"—actually getting the technology into the hands of those who need it and ensuring that they have the confidence and skills to make use of it.

The result is well documented: a divide between those with access to digital technologies and the confidence and skills to use them, and those who do not.[17] This can vary by age, but it varies by other important factors as well, such as education and financial resources.[18]

Everyone has a need for support of some type with digital media and emerging technologies.[19] Even those on the younger end of the "older adults" age spectrum, even those working in fields such as computer science: everyone finds the need to ask for help at some point from someone. This universal need among older adults connects to the needs of every generation. No one is born with the innate ability to handle all digital media—stereotypes about today's youth notwithstanding—even if it might sometimes seem that way. Everyone needs help along the way with digital media, and it is important that we understand how to make it accessible in diverse ways.

A way to think about these many aspects of support together: consider the many ways in which social context matters.[20] If friends and family encourage use of digital media and provide easily accessible, positive forms of support, an older adult is much more likely to use the tools and services. When they have even just a little confidence based on this social support, an older adult is much more likely to develop skills on their own, learn how to learn, and learn how to access support when they cannot manage on their own.

How do we help those who do not have access, for any number of reasons, to benefit from the positive aspects of digital technologies? The answer is a mixture of things. A series of key variables can make a big difference. For starters, social context matters. An older adult who has a spouse or partner in their life likely will have different needs than someone living on their own. Those who are closer to working age will have different needs than those who may never have worked using technologies at all or who are much older. Those who have access to a terrific library or who can pay for support services may have different needs than those who are isolated and less well-off financially. The strategy for ensuring support for older adults in their technology usage will have to vary based on these factors. But in every case, the "last mile"—or the way in which the older adult interacts with the technologies—matters the most.

Given that there is no one form of support that will make the difference for every older adult, it is essential to create accessible pathways to a variety of forms of support. These pathways cover that "last mile." In general, older adults prefer to get support from those they know and trust, so it makes sense to start there when possible. Some older adults may prefer to give and get support from one another rather than turning to younger people for help, but that depends a lot on the circumstances of the person's age, past training, and social context. If the social context does not easily allow for this trusted type of close support, many older adults benefit from support via formal computer classes at libraries, informal drop-in sessions at community centers, or for-pay offerings from stores.[21]

A lot can ride on an older adult having access to good support for using digital media. There are substantial social benefits to be gained from digital media usage in families and in friend groups, particularly for older adults. The use of digital media can be linked to healthy aging as well as the possibility of engagement in civic and political life, access to entertainment, and connection to loved ones at a great distance, topics we discuss in detail in future chapters.

A way to build a more equitable and just society is to ensure that older adults have access to the support they need to access digital media. As more and more of our systems and services become digitally mediated—for getting food, for voting, for paying for services, for health care appointments, for communicating with one another, the list goes on—we cannot forget that not everyone will find these new modes of access to be convenient or easy. Many older adults will be able to manage, but others will have a much harder time. It is essential that we do not leave them behind as technological change continues to accelerate. We also know that companies will simply not do this on their own—we have decades of evidence that "leaving it to the market" in technology leaves major swaths of people out of the loop and/or at substantial risk of harm. As the new wave of AI (artificial intelligence) technologies hits the market, we need to bear this lesson in mind. Some of the most critical areas where support can be crucial concerns security and privacy, the topics of the next two chapters.

4

Safety and Security

THE GREATER THE AGE, THE EASIER THE TARGET?

Owen was well into his eighties. He enjoyed going online every day to connect with far-flung friends and relatives. Mostly confined to his bed and his couch due to ill health, he was a heavy user of Facebook among other social media. He had moved around a lot, spending much of his career in Europe but returning home to the American South for his retirement. He also sent and received a lot of emails to stay in touch with friends and family, near and far.

His friends were always glad to hear from him, but some of his messages could be pretty confusing. It was not infrequent for Owen's email correspondents to receive an odd message from him:

> Dear John:
> I'm traveling in Europe and I've run out of money!!! I can't get a train ticket to visit my cousin because someone stole my wallet on the last train. Can you send some money FAST to my bank account to help me get home??? I really need your help or I'm afraid I will be in trouble. I will pay you back as soon as I get home. I promise.
> Your friend,
> Owen

The messages from Owen to his friends sometimes had curious spellings. Owen's friends would often call him on the phone instead of sending him money, and they would find he was home sitting on his couch and watching TV. Owen did not need any money for the train.

On Facebook, Owen's "friend requests" were constantly hitting the inboxes of those in his social circles. After the third or fourth request they received (they always said yes), his friends would call him up instead of clicking on yet another friend request. Owen would check his account and let them know they were already friends and no, he had not asked them to approve a new friend request. It turned out those requests were coming from fake accounts pretending to be Owen with the intent of harvesting information from those in his circles and developing a legitimate-looking account to then perpetrate scams.

Owen also spent a lot of time on message boards for his investments. He particularly favored the Bogleheads forums, established in honor of the famous investor Jack Bogle, who cofounded the Vanguard Group. After Owen received a suspicious message, he posted online:

> Just recently I got an email from "Vanguard" to my same email address saying, "Vanguard Investments - Important information about your new account, etc.," BUT it had a different person's name on the account. One I had never heard of before. That email did not mention anything about the type of account and didn't have any other info that was pertinent to me.

Respondents on the message boards told him not to click on any links, not to respond to the email, and to give Vanguard a phone call. Owen took the advice.

Owen also got a lot of text messages from people he did

not know. One recent string started with a photo of a person's hand holding out a nice bottle of wine—a 2017 Opus One. It looked like someone was standing in a wine shop, seeing if he'd like the bottle.

OWEN: Looks nice. Who is this???

OTHER PERSON: Oh is this not Elsa's phone? Sorry. The secretary must have put in the wrong name.

OWEN: NO this is not Elsa this is Owen.

OTHER PERSON: You seem really nice. Im Debbie.

OWEN: Hi Debbie. Thank you. You seem really nice too.

OTHER PERSON: I run my family's medical practice. What do you do, Owen?

OWEN: Well I am a retired business man.

OTHER PERSON: Its nice to meet you, Owen the Retired Business Man. Mind if I text you again some time? Gotta run! xx Debbie

Owen had amassed a lot of correspondents. Not all of them had his best interests at heart. One way scammers get into people's good graces is to pretend to contact them by accident to then see whether they respond, which opens the door for building a trusted relationship before they scam the individual out of money that can amount to considerable loss.

No one is immune from the security risks of using digital media. Identity theft remains one of the most common crimes in many nations of the world, affecting digital media users of all ages. Identity thieves target older adults in particular: compared to younger tech users, many older adults tend to have more money to steal and are perceived to be less technically savvy or to have decreased cognitive function.

The Federal Bureau of Investigation has been raising the alarm about a spike in online fraud against older adults such as Owen. In its Elder Fraud Report for 2021,[1] the FBI reported that over 92,000 older adults lost some $1.7 billion, a 74% increase over the previous year.

When asked why he robbed banks, the famous twentieth-century criminal Willie Sutton is purported to have said "because that's where the money is."[2] These days, older adults are perceived to be "where the money is." Their perceived or actual wealth is just one reason older adults are at a growing risk of fraudulent attacks online. Owen had substantial retirement funds, and he loved going online to research the markets, chat with others about his investments, and make trades—all of which could make him look like a great big target for scammers. Authorities believe that older adults are targeted more because of their potential to be in cognitive decline, due to an assumption about lower digital savvy, and because they may be lonely and thus vulnerable to an approach that involves a proposed romance or friendship, drawn into a relationship that is not at all what it appears to be.

It is true that older adults hold a large share of the wealth in many countries. In the United States, for instance, a recent study showed that 83% of the wealth in the country was held by people over fifty years of age.[3] Older adults are a target for scams in part because, like Owen, they are wealth holders.

The average financial loss of the older adults who reported to the FBI was over $18,000.[4] Very often these sums are gone forever with little recourse for recovering the funds. The understandable frustration, sometimes despair, that older adults express when they have suffered these losses is painful to behold.

Gathering reliable information about who has experienced fraud is difficult for several reasons. First, for many, it is embarrassing to admit that they fell for a scam, and so they may hold back from reporting it to an official agency. Similarly, they may not own up to such an experience when filling out a survey. Additionally, some people may not realize that they *have* been victims of a scam and so they would not know to report it. In Owen's story, recall that his friends received several requests from a Facebook account appearing to be his. Those who did not realize it was not Owen reaching out would not say that they have received messages on social media impersonating others.

With that in mind, it is interesting to see how older adults' experiences with different types of scams compare. Given the above difficulties with measurement, these are all likely to be conservative figures as they mask the experiences of those too ashamed to admit being scammed and those who were scammed without their knowledge. Among 2,000 older adults taking our survey, 72% reported having received an email pretending to be a company, the most common type of fraudulent activity we asked about. These can be dangerous as they often concern phishing attacks. In such cases, the fraudulent sender tries to gain access to the user's account information, including their password, by pretending to be an organization urgently needing the user to log in to verify some information. Those who click on the link in the email are usually directed to a site that looks like the organization but is in fact controlled by the fraudulent sender, who grabs the user's password as they attempt to log in.

Conveying a sense of urgency is one of the telltale signs of scams. Recall in Owen's example at the start of the chapter, the need for money was "FAST." The senders of such messages will prey on people's vulnerability in emergency situations and instruct immediate action to prevent the user from taking the time to reflect on the legitimacy of the message.

Another increasingly popular form of scam message is

to send a bill and thank the recipient for having subscribed to a service. Alarmed that they had not authorized such a purchase, the user may respond to such a message, thereby confirming to the scammer that there is a live person at the other end of the message willing to engage in correspondence. It is thus best not to respond to any suspicious messages. Even something as innocuous as requesting to be removed from a list or clicking an unsubscribe link can signal to scammers that their message met a potential target, thereby triggering the next stages of the scam attack. (How to unsubscribe from anything then? If it is something you are certain you had once subscribed to, then double-check whether the sender is truly who the message says it is by looking at detailed information in the header of the email—the underlying email address in particular—before proceeding to click on the unsubscribe link. If it is not something you recall subscribing to then create a filter in your email program—usually under settings—to have the messages automatically moved to your trash folder.)

Social media impersonation has become quite common, with half of the survey participants who use such platforms reporting that they had experienced this. About a third of people had received an email with similar impersonation. A quarter of respondents are aware that they had a virus installed when they clicked on an attachment in an email they received. The consequences of this could be wide-ranging from a user's online activities being tracked to sensitive information being siphoned off their device. A fifth of people admitted to having fallen for a scam, 14% reported having had financial information stolen, and 10% reported their identity having been stolen. For how substantial and potentially debilitating the latter offenses are, these are considerable occurrences.

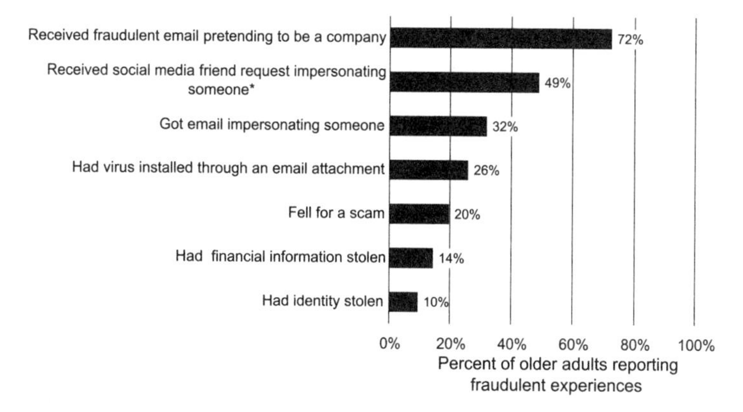

Prevalence of fraudulent experiences reported by 2,000 older adults (2003). (*Among users of social media.)

Many scams occur through personal messages. Over time these have changed in flavor and continue to be successful as a general form of fraud. There are the older types such as the ones where people impersonate a rich person from a foreign land claiming to need a way to rescue their money. More recent ones try to get more personal quickly. They often start out as an innocent-looking mistake by the scammer. They will say something generic like "Hi Tom, looking forward to seeing you tonight" or "Darlene, did you get my message?" to which the recipient replies by saying they do not know the sender. The sender then apologizes and starts a friendly chat often suggesting that they are from another country and sending the picture of a smiling young woman.

The recipient—whatever their age—may then start engaging in conversation with the sender (see Owen's example at the beginning of this chapter) and sooner or later gets swindled out of money. This can happen through suggestions for investments or through direct requests for help. Such messages can come through social media but are also common elsewhere, such as apps where people buy and sell items or even a phone's regular messaging app. As usual, if you do not know the person contacting you or if there is anything potentially suspicious about their language (spelling, style), it is best to move on. At minimum, it is important to get a second opinion. Another way to identify a scammer is to put their words into a search engine to see whether discussions of similar messages come up elsewhere.

Perhaps the most cunning is when tricksters reach out through dating apps since, by definition, the user is there to meet new people.[5] In such a case, the advice of being suspicious of an unknown individual is not helpful since the whole purpose of dating apps is to meet new people. It is nonetheless important to stay vigilant and scrutinize the details. Under no circumstances should a user of a dating app start sending financial resources to someone they met online and have never met in person. (Even meeting in person is no guarantee for avoiding fraud; for this we recommend watching the crime documentary *The Tinder Swindler*.)

There have been devastating cases of people losing their life savings to scams of this kind, where they are requested to send increasing amounts of money. The especially heartbreaking part of this particular scam is that it does not only have financial implications—and major ones at that—but people's emotions have been toyed with when they find out that the romance they thought they had was all fake. It is thus important to be very vigilant and ask detailed questions, search online for additional information, and ask friends and family for a gut check. And if there are not readily available friends and family, then a question to a librarian or someone at the community center could also provide needed guidance. For more background on this particular scamming practice, we recommend the "Pig Butchering Scams" episode on *Last Week Tonight with John Oliver*.[6]

There is a perception that older adults are more vulnerable than younger people. It turns out that research does not support the popular view of older people as universally gullible and easily tricked out of their money. Reviews of the research literature across several national contexts find that susceptibility to consumer fraud is no more common among older adults than among younger ones.[7] In fact, it is more common for adults aged twenty to twenty-nine to report losing money to scams (44%) compared to adults aged seventy to seventy-nine (25%). In cases where older adults are more likely to have fallen victim to fraud, it is likely because they are more often targeted by scammers. And if they do lose money, they tend to lose more money than younger adults.

Older adults who have not yet suffered cognitive decline have some advantages with respect to assessing information (we discuss this in more detail in the misinformation chapter). The experience that comes with age and is often coupled by heightened skepticism means that older adults may not fall as quickly for scams just as they can manage to resist some fake headlines and other misinformation. Older adults are often able to apply their longer life experience to identifying potential scams and not enter into trusting relationships with those who wish to do them harm.

One way users can avoid unwanted access to their accounts is through two-factor authentication (2FA), also known as two-step verification. This safety approach requires that in addition to typing in a password, the user also supply additional information by way of a code emailed or texted to them. Authentication can also happen through a dedicated app such as one called Duo or another called Authenticator. In 2021, Google started automatically enrolling 150 million of its users into such a login process and found a 50% decrease in compromised accounts.[8] While clearly an effective way to keep accounts safe, their uptake tends to be low when left up to users. In 2021, X (formerly Twitter) reported that just 2.6% of its active accounts had at least one 2FA method enabled. One experimental study enabling 2FA on Facebook found that older users were more likely to enroll in it in response to prompts, as were more active users.[9]

When we simply presume that older adults are more vulnerable to online scams, we capture only a part of the real picture. First, that statement is not universally true; while not comprehensive, the data seem to suggest that older adults are not per se more gullible or susceptible to online harms than young people. While there have been a number of studies of older adults and the security risks they run and how to address them, these studies have been fairly small scale and not comprehensive, suggesting the need for more research.

Older adults are an increasingly diverse group. Those in their early sixties who have worked in a technology-rich field are almost certainly much better able to navigate online safety issues than, say, a young person just entering the workplace who has never had serious computer training. Yet those who have been caregivers at home for their whole career, or who have worked with their hands as a manual laborer, or who have retired long ago may be at a significant disadvantage in terms of their ability to understand the workings of the technology and how a person seeking to do them harm might use the tools to trick them.

An especially relevant way that some older adults run the risk of encountering fraud is cognitive decline. Generally speaking, adults over sixty are more likely than younger people to suffer from cognitive decline, leaving them at greater risk of scams than their younger counterparts. The incidence of cognitive decline only grows from there, with research suggesting that as many as half of those in their eighties are likely to have either a diagnosis of dementia or "cognitive impairment without dementia."[10] Cognitive decline is the most likely issue leading older adults to be more vulnerable to poor decisions related to their finances, including vulnerability to financial fraud. There is a clear link between the greater likelihood of cognitive decline among older adults and their ability to make good financial decisions, including with respect to finance-oriented scams. The peak age at which we make good financial decisions, by the way? Researchers have pegged it at age fifty-three.[11]

For caretakers of those with significant cognitive decline such as Alzheimer's disease, there may be a need for being more hands-on than is otherwise advisable. From the most basic actions like how to retrieve a text message or call a loved one, depending on the stage of dementia, the older adult may need more involved protection against scammers. While "parental controls" are usually thought of as a way to keep an eye on young children's device activities, they may also be helpful for keeping older adults safe. There are implications here for privacy, of course, which we will take on in the next chapter.

One tiny study shed further light on this connection between cognitive decline and online security risks for older adults. In a study involving twelve interviews of older adults with mild cognitive impairment (MCI), every one of the families interviewed shared at least one recent incident involving safety and security concerns related to online activities. The good news? These families also all reported a somewhat informed understanding of the potential pitfalls of their older relatives' digital activities.[12]

For those who are older, a lack of awareness of how the technology actually works can be a problem that intersects with cognitive decline. This issue pertains especially to those older adults who retired a long time ago or who were not employed in an office-based workplace. Studies suggest that a declining tech savviness as older adults retire from the workforce can affect some of them. Social isolation can also reduce the availability of the peer support needed to provide assistance to older adults as they seek to navigate online environments,[13] which is why we spend a full chapter on discussing the importance of providing support. Some older adults have never been in the workplace during the digital age (or ever) where some of the latest scams are discussed and approaches shared for how to avoid them.

In workplaces where staff use computers as part of their jobs, it is typical to receive training in technology usage regularly. One of the most commonplace trainings focuses on how to avoid falling for scams and related online abuses. These trainings are extremely useful in sharing, for instance, how a fraudster might think about the effort to get someone to part with their personal information. Since these tricks change quickly, the trainings in many workplaces happen at least annually. In our workplaces, we regularly undertake mandatory trainings on technical safety measures to stay up to date on the latest scams and how to avoid them.

Older adults who are still employed or who worked more recently in computer-based jobs will have had more opportunities for pattern recognition based on training programs than those who have not. One special advantage of the workplace for security issues is learning from negative experiences, either one's own or others'.[14] Fewer people may be left behind due to a lack of digital experiences at their workplace as more and more older adults retire from jobs that have a digital component, training, and experiences. But in the meantime, especially those older adults who have not had the benefit of a workplace with digital training and experience continue to fall on the at-risk side of this digital security divide.

As the FBI and others have demonstrated, older adults face a growing onslaught of efforts to defraud them. Technology is often the medium used to commit that fraud. These scammers have moved their frauds from in-person attacks to digital attacks because they can reach more potential victims that way. Yet other frauds continue to be perpetrated by mail and other media. One example: a letter sent to an older adult telling them that they have won the Publishers Clearing House sweepstakes. They just have to offer their bank details and a small initial withdrawal to claim the big prize! Owen and his peers are the targets of scams on a daily basis on nearly any imaginable medium. Those scams are only growing in sophistication as the technology improves. A scary thought: greater adoption of artificial intelligence tools, well underway today, means that scammers have the ability to target and personalize their scams even more precisely than they have to date.[15]

Another reason why those now or recently in the workplace may be better equipped to deal with safety issues is that organizations often roll out required security features before individuals may come across them in their personal lives. Such is the case of two-factor authentication. Once a user gets nudged to use it at work, that experience can grease the wheels for its adoption for their private internet usage as well.

The scientific literature on older adults and online security is in its early days. The studies that exist are typically fairly small scale, involving a few dozen participants in a single country. These studies give some insights into how older people think about their exposure to online security risks, but the picture is still emerging—and experiences are incredibly diverse among older adults, as we have seen throughout this book.

Older adults seem to be well aware of the risk of scams online. One study highlighted the fact that many older adults were highly aware of scams and knew that those who targeted them were resourceful and tricky. Respondents in this study were highly critical of those who fell for scams, using terms like "gullible," "greedy," and "stupid" to describe their peer group members who failed to avoid the attacks. The principal coping mechanism reported among older adults in this study was "skepticism" and "common sense"—an approach that other studies have shown older adults to deploy to protect themselves from other online harms, such as mis- and disinformation.[16]

A similar study showed that a majority of respondents were aware of fraud, scams, and identity theft. These respondents also demonstrated a fear of their online personal information being disclosed in such a way as to cause reputational damage, humiliation, or embarrassment.[17]

There is some evidence to suggest that older women may be at greater risk than men for some types of online security attacks. A study of phishing attacks, in which emails are sent to users to try to coax them into disclosing personal information, showed that more than 40% of the participants in the study clicked on at least one such phishing email during the twenty-one-day study period. In this particular study, the authors concluded that older women in the cohort of those studied were the most susceptible group to phishing attacks.[18] The study cited earlier about enabling two-factor authentication on Facebook found that women were less likely to take this extra step, which may explain some of the higher prevalence for falling victim to phishing attacks.[19]

Older adults typically tell researchers similar things about the approaches they use to avoid scams and other security threats online. These approaches employed by older adults are essentially identical to those used by younger adults. In one study, the mitigation strategies reported included configuring privacy and authentication settings, installing protective security software and services, and deleting or refusing to provide personal information when asked by those they did not know.[20] In a similar study, older adults reported positive benefits from strategies, including risk aversion, security education, and perceived ability in finding information online.[21]

In our survey, we asked older adults whether fear of their identity being stolen held them back from signing up for or downloading certain online services and apps, and over half (54%) said that to be the case. Fewer people reported fear of their money being stolen (40%) as a reason to hold back from adopting new services. Across the two concerns, 59% said that one or the other reason kept them from trying out a new service. Given all that older adults can gain from using technologies, it is important not to have them skip such opportunities due to security fears, however justifiable. We note again the importance of assistance that can help them address these concerns so that they use technologies in an informed and safe way.

Pushing back on the powerful negative trend—the rising tide of security attacks on older adults and how this scares people from adopting helpful services—is a major challenge. Law enforcement officials, despite their best efforts, have had a hard time stemming the rising tide of security attacks on older adults. The solution is unlikely simply to ask our law enforcement services to "do more."

Experts agree that education of older adults to inoculate

them against these harms is the best approach. Experts also agree that this education is hard to do well and at scale.[22] Community-based training and training in the home, conducted by trusted parties, are the best approaches and most likely to be successful. This education is probably best done one on one, either offered by a younger adult who knows the material well or a community member or partner who has been equally well trained. This guidance and advice is most likely to succeed when offered up front, when an older adult is just starting to use a new service or technology.

The strongest piece of advice we can offer: invest the time at the outset to set things up well. Put in the time and energy to make sure you or the older adult you are helping understands how to use the technology and can ask all the questions they may have. This upfront investment is likely to pay dividends down the road. Older adults seem to ask for advice early on as they are getting started, no surprise, and that can help with accuracy of responses to security-related questions. They are most open to hearing advice, and they internalize it best, at the outset. We mention this in the last chapter, too, but if you remember just one thing from this chapter: the research shows that advice is most effective right as older adults get started with a particular technology or device.[23] And it is most helpful when delivered patiently and respectfully.

The research on older adults and security points to the notion of control as a key concept for how to help increase security and reduce vulnerability. If the older adult feels confident and in control from the outset, they are more likely to get the most out of the experience. Where the opposite is the case—when they feel out of control—they are likely to give up.

For instance, older adults reported thinking a lot about who had "control" over their passwords and other identifying information. This concern over losing control leads some older adults not to use technologies that can be helpful in terms of reducing security risk, such as password managers.[24]

Password managers are considered an effective means of increasing security in everyday technology usage. The frustration is one that any user of digital technologies knows well. The options for how to keep a long list of online passwords are lousy, typically: using a single, simple, memorable password for everything is easily cracked or found. Leaving a single sheet of paper in a drawer is likewise easy for someone to steal—and they then have a road map to everything you own (in digital form, anyway). An online spreadsheet with all passwords may not be well secured.

Security experts report that using biometric technologies such as fingerprint and face recognition, combined with a secure password manager, multifactor authentication, and so forth can substantially improve security and privacy. Some of the most effective tools, such as two-factor authentication and password managers, can prove challenging and frustrating for older adults new to the technology. The failure to use these tools can put older adults at greater risk, while getting them comfortable can be intensive and time-consuming. But given the number of problems that can ensue if good protection measures are not in place, patient assistance up front offers the highest payoffs later.

The venue for training older adults in digital security can vary based on the needs of those who are doing the learning. If we could recommend anything, it would be customized, one-on-one training done in homes, in person, by knowledgeable and trusted trainers. Small-scale studies do show benefits of training older adults—comfort levels rise, and the older adults are in turn more likely to use tools that could be helpful to other older adults in their communities.[25] Researchers also point to the notion of proximity. "Close social help" is most effective in training older adults in security and privacy.

The best news of all may be that older adults appear highly motivated to learn. Studies have shown that older adults often do intend to learn and explore new technologies—if they can get comfortable and have a sense of control.[26] Older adults welcome support from others they trust.[27]

This positive news is often then translated into efforts to set up home-based training approaches that build on this intent to learn. In technology-rich middle-income urban Indian homes, for instance, studies have shown that "self-appointed tech managers"—a family member who is knowledgeable about security and privacy issues—can be essential to safe and secure technology usage throughout the extended family, including older adults. Heavy-handed approaches by these family-based technology trainers tend not to work; light-touch, supportive approaches are much more likely to succeed.[28]

Follow-through by family members is essential. Even if there is a good start when the older adults begin using the new technology, too often relatives are off to the next thing. Family members mean well but often do not carry through in helping their older relatives with online security training and so forth.[29]

Even if the home appears to be the best place for learning and teaching about digital security, it is not always possible. The good news is that community support is also effective. Successful training and learning can occur in senior centers, libraries, and other community gathering spaces. The training can be carried out in person and virtually. The train-the-trainer model is an approach whereby people who get trained then become trainers themselves and can themselves train others to be helpers. Such models can scale and can make a big difference in terms of peer learning among older adults in communities, especially when younger family members are not available, not skilled, or fail to follow through.

In considering approaches to community-based training, the research points the way to what can be successful. Older adults have the capacity to support one another when it comes to digital security management. Studies have shown that community facilitators—often older people themselves—can play a positive training role, especially in residential communities. Once again, the notion of self-efficacy and control show up as key factors. Where older adults believe they are in control and have the skills they need, they build confidence, use the technology, continue to learn, and share this knowledge throughout their communities. It can be a virtual cycle: the success and confidence of individuals can help lead to community strength among older adults.[30]

The research about older adults and security is clear: the issues facing older adults are quite similar to the issues facing younger people, with the possible exception of additional cognitive decline such as memory problems and language difficulties. What that means is that many of the approaches that work well in places of employment, training centers, and libraries for the population at large can be adapted for older adults as well.

A few attributes that appear in trainings in workplace settings can help with those at home or in community settings. Interactive trainings work better than lectures. Exposing older adults to the types of threats they face, using specific examples and narratives, is essential to success. The stories described should be relatable and set in context for older adults. Trainings need to be both repetitive and done over time, while also being differentiated so as not to result in boredom and lack of focus.

The strategy of "prebunking" that we also take on in the misinformation chapter can be helpful too: by making the older adult aware in advance of what they might experience, they will be more ready to address it or take precautionary measures. A good example comes from banking. Many banks inform their customers online and in person about scams that are regularly affecting their client base. This strategy works for all ages; for older adults, it may work especially well by tapping into their ability to use pattern recognition over a long lifetime. Quick one- to two-minute videos on YouTube and TikTok can explain the basic modus operandi of typical scams using visuals. For example, in one such video by the Zelle banking app, the voiceover describes how a text message pretending to come from one's bank can swindle the user out of money.[31] The narrator explains that any response to such a message will prompt a call from the scammer, which will then lead the unassuming user to give up access to their bank account. This video is just 81 seconds long and conveys the most important features of such a scam attempt.

Another way to impart details about scammers is through a humorous angle. Lots of videos exist on YouTube, TikTok, and the streaming platform Twitch that show how knowledgeable and responsible users flip the script on scammers to toy with them, frustrate them, and in some cases go so far as to expose them. One such popular account is Kitboga (a pseudonym to protect the creator of this content) who shows in detail what it is like to interact with fraudsters trying to scam him out of his money.[32] His videos are entertaining, but they also have an important educational function in that they show the viewer how criminals operate. While some of the scammers are the stereotypical lone wolves, others are seen in call centers defrauding unsuspecting victims in organized ways, showing just how big of an industry this is.

While scams can come in countless forms, most follow popular themes that are worth discussing among family and friends. Some popular ones include a message conveying an issue with package delivery; pretending to be a utility company with an urgent message or unpaid bill; or sending the recipient a fake bill thanking them for payment to prompt concern and a response contesting the (fake) charge, which then results in communication that opens up the user to actual fraudulent charges.[33]

The net effect for some older people is that they then do not use digital technologies at all because they are afraid of this type of scam. While rational, fears about security and privacy in turn can mean that such people do not get the good parts of the technologies, for instance, the possible well-being effects for their health and social lives. That makes these scams especially pernicious: not only do the scams put older adults at risk of direct financial harms, they also rob these people indirectly of other benefits out of fear. Concerns about online security may be leading some older adults not to take advantage of services that may improve their health.

Multiple studies have focused on the reasons why older adults would or would not adopt online health management tools. The main barriers? A simple lack of knowledge about these technologies and concerns about privacy and security.[34] As other studies have shown, the fear of insecurity—and lack of control—leads some older adults not to take advantage of digital tools that can serve their well-being and health. Another small study interviewed adults after their participation in a training program.[35] The average age of respondents was sixty-four and they were half white, half African American. Some of the concerns were off base, demonstrating a lack of basic understanding of how the internet worked, combined with a strongly negative view of cybersecurity. Some of these respondents reported that they refrained from using potentially helpful online tools due to these fears.

Flipping this script seems like the most promising angle. If older adults had greater knowledge, skills, and confidence—a greater sense of control—they might well then have a greater comfort level with using new technologies. This confidence could in turn lead to greater health and social outcomes for older adults. This sea change may be especially crucial for lower-income adults.[36]

It might be tempting to look to the technology itself to solve these threats. Online and at-scale efforts so far have not proven to be a panacea or even particularly successful. Efforts by institutions such as banks to reach customers and raise awareness hold some promise—but only to put the issue on the radar screen for older adults as part of a "prebunking" strategy. These online, at-a-distance strategies are unlikely to draw many into an actual training or an in-depth, persistent effort that will enable them to feel and keep themselves safer online.

Another approach that plainly does not work so well: leaving older adults to their own devices (pun intended). The research makes this plain. Older adults tell researchers that they are not comfortable using the internet to search for security information or for approaches to troubleshoot security issues they might face. Older adults report that they do not seek out cybersecurity information themselves and typically learn about these matters from friends and family members.[37] Another study has demonstrated the limitation of "just telling older adults" to use advanced technologies, such as two-factor authentication. Without the support from caregivers or peers, they are unlikely to take up these tools on their own, as helpful as they may prove to be.[38] Those with the skills and the knowledge have to offer personalized help when possible, early and often.

Stories are memorable.[39] No matter how rich and compelling the statistical data we collect and present, no matter how rigorous or repeatable the methodology we employ, these facts are likely to fade quickly from a reader's mind. A story is much more likely to linger, which is why we have shared several in this chapter and point you to a few others here:

- "The day I put $50,000 in a shoe box and handed it to a stranger, I never thought I was the kind of person to fall for a scam" is the first-person account of how an educated financial-advice columnist handed over $50,000 cash to scammers.[40]
- "Finance worker pays out $25 million after video call with deepfake 'chief financial officer'" reports on how a worker was convinced to send major funds to criminals who staged a video conference call where everybody but the victim was fake.[41]
- "Cruel way scammers stole dad's life savings" tells the story of a man getting scammed out of his life savings inspired by an email of promised inheritance from his homeland where he had many relatives, making the inheritance idea seem plausible.[42]
- "A 74-year-old man lost $50,000 life savings after downloading an app to order Peking duck" describes how a Facebook ad prompted a man to download an app, which then required a payment through another app, which then crashed the victim's phone while wiping out his bank account.[43]

As these cases demonstrate, people of different backgrounds—and ages—can lose significant money to fraud. Beside the significant financial hardships that victims experience comes the social embarrassment of having fallen for scams. The truism that stories are memorable is at work for older adults and their safety too.[44] Researchers have shown that older adults often find out about security and privacy issues through stories, so it is important to share experiences, whether our own or those we hear about. Learning about what can go wrong attracts attention, helping us be better prepared to meet the rising tide of scams and frauds both online and off.[45]

A close cousin of security is privacy. In the next chapter, we look at effective ways of addressing privacy concerns to make sure that they do not deter older adults from adopting important life-enhancing technologies.

5

Privacy

WHAT'S WORTH THE PRICE OF PERSONAL DATA?

Helen's family had a decision to make—a big decision.

Helen had reached her nineties. She was in wonderful health overall. Her lifetime of good friendships, strong family connections, and healthy lifestyle choices were paying off. Her doubles tennis play may have morphed into pickleball in recent years—the smaller court was much easier on her knees—but she still got out on the court from time to time. She enjoyed her days and loved the family home in which she continued to live. She still enjoyed working on her beautiful garden. She occasionally had helpers come to check on her and help with some tasks, but money was tight, and Helen felt that a full-time caregiver was out of the question.

Her children had thrived, all enjoying fulfilling lives, friendships, and careers. One of the three had children of their own; the pictures of these grandchildren flashed up on Helen's smartphone screen to her eternal delight. However, all three of her children lived far from the US Southeast, Helen's home. In fact, none of them lived in the same country as she did. They all visited regularly and communicated nearly every day by phone and FaceTime, but they were not close by. And that was unlikely to change anytime soon since each one had a full life and enjoyed their respective communities too.

Helen's three adult children had gathered to celebrate her ninetieth birthday. Everyone assembled knew the topic they needed to discuss: was it time for their mother to sell the house and move into an elder care facility? As anticipated, Helen did not want anything to do with the idea. She loved her life in the home, and, as she told her children, her time in the garden sustained her on many levels. It kept her body and mind active, it reminded her of what she cared about most, and she saw no reason to give it up at this time. The children were reasonably concerned about her health, she contended, but there was actually no cause for worry. Her primary care physician had assured her that she was as healthy a ninety-year-old as he had ever seen.

Helen's children had come prepared for this deft move by their mother. If she were to stay in the home longer, she needed to consent to a few things. First, they wanted to install a few cameras so they and perhaps a nurse or nursing service could keep track of what was going on in the home. They wanted to add motion sensors in parts of the home not covered by the cameras. They wanted her to carry a device around her neck that would allow her to click a button if she was in pain or worried about something, indicating to an on-call nursing service that she needed help. And they wanted to explore a device that she could wear on her wrist that would monitor her vitals or perhaps a heart-rate monitor worn on the chest, sending data back to a cloud-based server that would be cross-checked continuously against her previous vitals and the standard levels that health care providers wanted to see in a healthy person of her age.

Her children wanted to know: would Helen consider these steps to ensure her health while she remained on her own in the comfort of the family home? They knew that it would detract from her autonomy and sense of privacy. It would require her to submit to various kinds of surveillance that would be completely new to her. She would have to get used to a sense of being watched, of being monitored, at all times. But these trade-offs would come with a clear benefit: her children felt it could keep her in her own home indefinitely, at least so long as her general health looked good.

Pleased that they had not pushed on the sale of the house but with unease about her privacy, she thanked them for their concern. Helen concluded the discussion by saying simply that she needed to think about it. After just one family discussion, they were not going to resolve such a major life decision that day. The children were not surprised.

Helen watched from her front living room window as her children left her driveway on their way to their respective destinations across the globe. With a chuckle, Helen wondered to herself if they would start on her next about her driving, which mercifully had not come up that day—and then quickly realized there might be an app to track her driving too.

James is in his seventies and retired just a few years ago after enjoying many years at the business he had built from scratch. He had a very able assistant who took care of most logistical matters, including the technology setup at the office. This was convenient at the time since it gave James more headspace to focus on his company's core mission. But once he closed shop, he realized that he had less technical know-how than would be ideal for his retirement years. He felt like he had to play catch-up when assistance was no longer a few feet away.

As a widower living on his own in a big house on a small midwestern town's cul-de-sac, James has gotten used to his privacy. His two adult children moved to distant states years ago. His extended family also lives far away, so he is used to relying on himself to get things done. James is active in his church, volunteers, and stays social by regularly meeting up with local friends for lunch and by seeing far-away friends on trips they take together.

James is financially comfortable and is thus able to pay for assistance, whether that concerns getting his lawn mowed or hiring someone to install a new printer. And while he is not timid and readily asks for such help, he quickly gets overwhelmed by lists of tasks, so constantly calling a technology specialist is not realistic.

James has embraced some aspects of technology but is highly skeptical of others. He is fiercely independent and private, so the idea of having gadgets everywhere around his home collecting information about him not only holds no appeal but is downright disturbing. Through a younger tech-savvy friend, years ago James learned about the search engine DuckDuckGo, which does not track personal information. He immediately switched to it for his daily searches to guard his online privacy and has been happily using it ever since. He is hesitant to sign up for too many online services—whether that concerns delivery apps or social media—as he wants to avoid leaving lots of digital traces behind. Accordingly, when his younger tech-savvy friend suggests services that would make his life easier, if they require a sign-up—which most do—he shies away from embracing them. This limits the extent to which James can thrive in the digital age.

Both Helen's and James's stories are ultimately about choices and trade-offs. There is the risk of harm to the older person when they are alone compared to the loss in privacy if digital devices track their activities. In both cases, these healthy older adults put a premium on continuing to live in their private homes and in many ways are doing well in this setup. In both cases, they have things to gain from technologies. Helen could benefit from getting any immediate health issues addressed quickly. James could profit from the convenience of online and app-based services such as borrowing e-books from the local library or keeping track of his to-do list through helpful task managers.

Digital health tools can offer life-saving assistance, but a number of people could have access to information about Helen's every move and many data points regarding her health; even videographic evidence of how she spent her entire nineties would be potentially available for consumption. These digital health and safety technologies could both help make her safer and undermine an aspect of her well-being: her sense of personal data privacy.

Similar to most older adults, Helen has a "degree of concern" about her data privacy. It was not her number one issue, and she was not so concerned about it that she rejected out of hand the ideas her children brought forward. In fact, she thought some of the ideas were pretty good—she might even feel a greater sense of personal safety knowing that she could just press a button and have someone come help her. But still, the idea of giving up so much privacy? Was that really necessary?

Older adults tell researchers that their concerns about data privacy can range from mild to deeply worried (that would be James's case).[1] In most studies, the largest number of older adults fall somewhere in the middle of that range. As with many complex topics, older adults demonstrate consistently that they hold a wide range of views about personal data and privacy—sometimes inconsistent views, in fact. One thing is for sure, though: relatively few older adults hold the most extreme views about data privacy. In other words, most older adults neither express no concern about data privacy nor express such deep concern about it that they are immobilized by it. In most studies, most older adults fall somewhere in the middle of the spectrum: expressing some concern about data privacy but not so much that they want to stay completely off the grid. Terms that researchers use to describe this large group of older adults include "relaxed pragmatists" or "medium concerned," depending on the language of the study.[2] What is most important for Helen's story: these concerns about data privacy can sometimes be significant enough to prevent older adults from using specific digital technologies that might be useful to them and contribute to their well-being.[3]

Overall, older adults are using new digital technologies more and more as time passes. But one of the three main reasons that they might not get started with new digital technologies is their concern about data privacy. AARP's research arm, AARP Research, conducted an online survey of 2,807 adults aged eighteen and older to understand attitudes toward privacy. Between 2020 and 2021, the overall usage of digital technologies—from purchases of new devices to reported time on applications—increased. At the same time, this AARP Research survey showed that 34% of those aged fifty and older cited privacy concerns as a top reason why they might *not* adopt new technologies. The other top concerns they cited were cost (38%) and lack of knowledge (37%). More than 80% of those fifty and over in the same study told researchers that they were not confident that what they do online remains private.[4]

In the survey we ran in fall 2023, we found that close to half (47%) of older adults reported not having signed up for or downloaded certain online services or apps due to fear of their activities being tracked. About a fifth (19%) cited a fear of people in their social circles finding out too much information about them as a reason not to try a new service. Clearly, privacy issues are top of mind for many older adults.

Older adults tell researchers that they have a range of specific concerns related to their privacy in the digital arena. First, older adults worry that people, companies, and governments may get access to their personal information without their knowing (and, of course, older adults are quite right to wonder about this phenomenon in the digital age). Second, older adults worry that those same people and institutions might misuse their personal information—once again, a highly reasonable concern. Older adults rightly worry that people and institutions they do not know might practice surveillance over their usage of digital services. As we saw in the chapter on security, older adults are worried that their computers might be hacked or they might get drawn into a scam that will result in their personal data being released or misused somehow—and embarrassing them in the process. Every one of these concerns is valid.[5]

The notion of control is important in the privacy domain, just as we saw in the security domain. Older adults demonstrate a desire for control over how their personal information is used and seek to be active, not passive, in terms of their data usage.[6] This finding is consistent with the discoveries about older adults' worries about new technologies: they worry that other people may be using their superior knowledge of the technology somehow to track information about older adults and use it without their knowledge or permission. Again, these concerns are extremely reasonable given what we know today about how many companies and governments around the world have acted with respect to individual data privacy with little or no effective notice to individual users.[7]

Much of older adults' concern about data privacy represents a threat of the unknown and how that leaves older adults without the control they seek. Researchers who have drilled down in discussions with older adults find that they do not have a specific fear of specific people or companies taking or using their data, but they do have an acute fear of threats from unknown people or entities.[8] The lack of transparency in terms of who can access their private data and what they do with it is a big concern for many older adults—again, much like in the security domain.[9] This concern about what may happen with their data, and uncertainty about who might do something with it, may lead them to consider abandoning use of the technology[10] or, as we saw above, not even giving it a try in the first place. More than half of those fifty and over told AARP that the "potential risks of companies collecting data about me outweigh the benefits I get."[11]

Not all older adults have as much awareness as others do about privacy. Some studies show a "lack of privacy awareness" in general among older adults.[12] Many older adults lack a granular understanding of how data are used in the digital environment by those who collect them. They find much of the digital world to be "mysterious." Artificial intelligence appears to be especially mysterious for older adults—though it is safe to say that artificial intelligence is hard for almost anyone, save perhaps computer scientists and those who write algorithms for a living, to understand completely.[13]

As one example of this mysterious quality of new technologies, we asked participants in our survey a multiple-choice question about what ChatGPT is from among the following six options.

A feature that allows gamers to communicate with others on their team.

A messaging application that helps people communicate with their friends.
A type of conversation that helps users understand growing privacy trends.
A social media feature that helps people react well to their friends' posts.
A program that helps identify and fight against suspicious code on the computer.
A program that generates humanlike responses to questions.

A type of generative AI, ChatGPT is a program that generates humanlike responses to questions. Just over half (54%) of respondents got this right, something much more common among the youngest old in the study (60–64) at 64% compared to 36% of the oldest old (80 or over). There was also considerable difference by education: 42% of those with no more than a high school degree got this right compared to 69% of those with at least a college degree. We note again the diversity in skills that exists within mature users.

Many older adults worry that websites they visit ask for too much information about them.[14] Many older adults expect a "return of value" for sharing their data that they do not always think they get.[15] Older adults are often clear on the fact that their data are being collected by companies and used for various purposes, such as advertising. This knowledge leads older adults to think about whether and how they might consent to sharing their information. This is where older adults' lack of an extreme position on data privacy affects their decision-making and behavior: for many older adults, it is rarely a certain yes or a hard no when it comes to parting willingly with personal data in digital form.

Despite the widespread concern they have about data collection, a significant number of older adults are willing to "make a deal" to disclose their personal information. The online survey released by AARP Research in 2021 showed that 42% of adults aged sixty to sixty-nine would share their personal data with companies in exchange for "cash rewards." That number fell to just over 25% for loyalty rewards, coupons, discounts, or free products or services. For at least a quarter of older adults, it appears that it is possible to put a price on their private information—and that the price might be as low as a coupon for a sandwich or a free drink at a coffee shop. Perhaps it helps to know that 18% of the same group would give up their private data in exchange for a gift to the charity of their choice.[16]

Helen's children in the opening story of this chapter were in essence asking their mother to make another type of a deal. Absent a better approach, they wondered whether the family could rely on new digital technologies to give their mother a bit more physical safety, a better way to ensure her health and well-being—in exchange for a decline in her data privacy. If Helen spent her days puttering around the garden and listening to the radio in the kitchen, maybe making the occasional trip to the bakery in town, there would be little need for anyone to track any data about her whatsoever. The multiple solutions that her children proposed would sharply diminish her privacy. That said, the chance to remain in her own home, and for her children to worry less about her—well, that is worth a lot more than a coupon or loyalty points at a coffee shop.

The Pew Research Center asked adults of all ages about whether they find the use of personal data acceptable in various scenarios. Older adults' perspectives varied considerably from those of younger groups. Older adults were much less likely to find it acceptable for social media companies to monitor users' posts for signs of depression with the goal of identifying users at risk for self-harm so that they could be connected to counselors. On the other hand, older adults were much more likely to say that it was acceptable for DNA testing companies to share customers' genetic data with law enforcement to help solve crimes or for the government to collect data about all Americans to assess potential terrorist threats.[17]

Some of the data about how older adults act with respect to their data privacy compared to younger people are inconsistent. One research study suggested that older adults are less concerned about data privacy than other groups studied.[18] Other findings suggest the opposite: that older adults are more worried than, for instance, younger working adults about data privacy.[19] Another study indicated that older adults are less likely to share personal data and less likely to use the privacy tools on social media, such as Facebook, than younger users.[20] Yet another found that older adults are more likely to share publicly on Facebook.[21] Other studies find no age differences, for example, when it comes to specific privacy-protective behaviors.[22]

How might one reconcile these apparently conflicting findings? There are a few interpretations. If a study is too small or reports on a nonrepresentative sample, or a researcher overclaims based on the data they have collected, we might be led to reach too strong a conclusion about how the findings relate to all older adults. Another possibility is that each of the studies is accurate, but they have simply focused on older adults in particular subcommunities, or within a socioeconomic stratum, or a certain geography—any of which could affect the particular findings.

Academics are on strongest ground when working with empirical data collected based on sound, well-described research methods. These should be presented in a way that they can be replicated by others, although there may be exceptions to the feasibility of replication, such as when doing research during crises like COVID lockdowns or in the immediate aftermath of a terrorist attack.[23] That said—and we really believe it—that is not the only mode in which academics operate. Sometimes, absent lots of reliable data derived from replicable methods, academics also put forward theories that might or might not be true but that offer a potential way to understand a phenomenon. Theoretical claims therefore take on a different kind of quality—they are more of a hypothesis, an intriguing possibility which might or might not pan out. These theories operate also as an invitation to other academics to do further work to test the theory—to push the boundaries of our knowledge forward. This interplay works itself into a fascinating (for us, anyway!) discourse at conferences, seminars, online, and in journals that publish such work.

One theory that has intrigued many scholars, including us, is called the privacy paradox.[24]

One possible way to resolve the issue of the apparently conflicting findings about privacy attitudes and behaviors is the "privacy paradox."[25] This concept is not specific to older adults; it is a theory that may apply equally to adults of all ages. Nevertheless, this theory may offer a helpful perspective for making sense of the conflicting findings when it comes to older adults and their views on privacy.

The privacy paradox refers to the gap between the concerns people express about their privacy and how they behave. Users often share more content than one would expect from people who are purportedly concerned about their privacy. Thus the paradox: why share when there is such serious concern? Research has found that users still share because they do not see an alternative to existing in a digital world without making information about themselves available.[26] People describe feelings of apathy and cynicism, given that they believe that utmost data protection is only possible if they opt out of all digital services, not something realistic or desirable for most.

A poignant example of how hard it is to maintain one's privacy concerns the very personal information that a woman is pregnant, something that companies sometimes know even before the grandparents. An (in)famous case comes from the Target retail giant's marketing prowess in the early 2010s.[27] Because pregnancy is one of the rare occasions when people are especially impressionable for changes in shopping habits, the data team at Target put effort into figuring out when women may be pregnant to send them related coupons. A father once went into a Target angrily telling the manager that the store should not be sending his teenage daughter coupons for baby clothes as though she were pregnant. A few days later, he found out that his daughter was indeed pregnant.

Another explanation for the paradoxical results of data sharing despite privacy concerns is that it takes skills to be aware of and understand the many ways we leave digital traces behind and what we may be able to do to avoid this. Some people will recognize this in many situations while others will not. Some users will know that privacy settings can be changed while others will not. And some will know how to change the privacy settings to meet their preferences while others will not.

Sociologist Janet Vertesi took privacy protection to the max when pregnant with her first child. She did all that she could to avoid companies finding out that she was pregnant. Vertesi documented just how painstakingly careful she needed to be to avoid traces of her pregnancy getting into the hands of data hounds.[28] She avoided shopping online for anything having to do with her pregnancy and always used cash in stores. Some specialty products are not necessarily available at the local mall, however, so then she had to purchase gift cards with cash to use on Amazon. To avoid Amazon tracking her, she set up a separate account with an email address running on a personal server and had products delivered to a locker rather than her home. Her local store had a sign on the wall stating that the store "reserves the right to limit the daily amount of prepaid card purchases and has an obligation to report excessive transactions to the authorities." So even purchasing gift cards with cash was not a trivial undertaking.

Some of her family members sent her messages on social media to congratulate her on her pregnancy. She immediately deleted these threads confusing the senders who thought private messages were private. And while such private messages may not be visible to other users, they are certainly visible to the companies providing the tools. The way to think about this is not to imagine an employee of the messaging app company sitting at a desk reading through one's individual messages. Rather, it is to recognize that such messages are part of a database the company compiles about the user that it can then analyze in the context of the millions of messages it has from other users to find patterns across users. In that sense, the devil is not so much in the details but the large aggregation of the details.

For researching baby-related information, Vertesi turned to the privacy-preserving Tor browser. However, that service is often used for illicit activities, and so her increasing usage made her profile look more and more like that of a potential criminal. What does it say about the state of the digital world that trying to stay private about personal matters makes one look like a criminal?

As a researcher who studies technologies, Vertesi had the skills to pull off her goal of keeping her pregnancy away from watchful corporate eyes. She knew what could result in privacy breaches—a lot!—and she knew how to circumvent those actions (for the most part). Her case is an excellent illustration of why the privacy paradox is not actually much of one. To exist in the digital world, it is almost impossible to avoid sharing one's data. This is rarely thanks to choice; rather, it is a necessity. In a few pages, you will see our call to governments to nudge organizations to meet user needs better on this point.

Throughout this book, we examine approaches to how to assist older adults in navigating the increasingly complex digital world. As we consider how caregivers and loved ones can help older adults with respect to data privacy, we return to the framework of successful aging in our digital era. With enhanced education about what is really going on and with the transfer of skills, older adults can gain the confidence they need to be able to use new digital technologies in a way that is comfortable for them. Strategies need to put older people in positions of greater control over their data. The goal should be to empower older adults to understand the trade-offs they are making with their data. Ideally, in the end, they will make choices about when and how to use the new technologies that might give them the greatest benefits while avoiding those that represent a less worthwhile "deal."

The good news: the strategies for support that can help older adults are not all that mysterious. Most of the approaches that we propose in other chapters, such as the immediately preceding chapter on security, will work in the context of privacy. Younger people, family members of all ages, older peers, community members—all can be a part of the solution in terms of sharing knowledge and skills with older people. Programs at libraries, community centers, and elder care facilities all have shown promise in terms of educating and empowering older adults with respect to their data privacy.[29] Sometimes, companies go out of their way to try to empower older adults (among people of all ages) by sharing knowledge about how new technologies work, what data are being tracked and by whom, and so forth.

Many older adults are "heavily dependent on others" to help navigate their specific concerns about data privacy.[30] Younger relatives who have a loving connection to older adults and know their preferences are the ideal people to help and offer the most promising approach when it comes to privacy help and education.[31] As we explain in the chapters on support and security, family members show a willingness to help older adults in their lives but are less good at following through on that willingness.[32]

Local community institutions offer an array of programs for older adults to help them use computers and other digital devices more effectively. Libraries, community centers, and senior centers are often effective providers of these programs. When it comes to privacy, these educational efforts could focus on how to give older adults a greater understanding of what is going on when they share their data and thereby restore their sense of control to the greatest extent possible. Older adults emphasize over and over that a lack of transparency in terms of what is happening in the digital domain is part of what makes them reluctant to use technologies that may otherwise be helpful to them, socially or health-wise.[33]

Researchers suggest a few specific strategies that may be especially useful to teach older adults how to manage their data privacy. Recall the most basic level of skill that we have introduced already: awareness. It is important to make older adults aware that privacy settings often exist for various services where they can change the default, a default that is usually much more generous in how far and wide one's data traces will be shared than most users want.

Once the older adult realizes that they can personalize privacy settings, they may still need help in configuring these settings to reflect their own preferences. Talking through the various privacy options can be helpful, as can walking the older adult through the steps needed to alter the settings. That way, as alerts pop up or the user has a change of heart later on, ideally the older adult can make further modifications themselves without having to rely on others, who may not be available. Researchers point to how teaching older adults a series of tactics may help: how to use authentication tools, install protective software and services, delete information that could be captured by others, ignore social media requests, use pseudonyms, and refuse to provide personal information in the first place.[34] There are some specific areas where older adults may need precise guidance, especially with respect to the use of passwords, which tend to be very basic when created by an older adult without guidance.[35]

Alongside caregivers and relatives, companies have a major role to play—and they could do a lot more than they are doing today with respect to educating and caring about the data privacy of older adults. AARP research bears this out: its surveys show that older adults feel that companies need to do more to engender their trust. There are four areas where companies need to do much more. Companies (and governments for that matter) need to be much clearer about what exact data they collect, and they need to do this in easily accessible language rather than the legalese they employ when they do disclose their practices. Companies need to provide much greater transparency about what they (as institutions) do with older users' data, especially health data. Companies need to allow older adults much greater control, using much easier tools, over what is done with their personal data. And companies need to improve dramatically the ability for older adults to opt out of sharing data that they are not comfortable sharing. More than 90% of adults aged fifty and over believe that such steps are important for companies to take in order to encourage people to use their services.[36]

As with Helen at the start of this chapter, the concerns that older adults have about their personal data come home to rest particularly in the domain of data about their health. Technologies now exist that could significantly improve the likelihood that an older adult could remain living independently in the family home for a long period of time. Without some of these tools, however, it might only make sense for that older person to move to a place that can offer greater support. Many older adults would vastly prefer to live independently at home rather than moving to an elder care facility.

One of the tools that can make such independent living possible is the use of aged care monitoring devices (ACMDs). These devices can help keep older adults safe by tracking vital elements of their well-being and alerting others when they need help. These ACMDs, however, track an extraordinary amount of health-related data to accomplish these health and safety goals. Older adults told researchers that they have concerns about ACMDs along many dimensions: ongoing and unwanted surveillance, the use of data by people and institutions that the older adult does not know, breach of confidentiality, disclosure of information to third parties, interference with the decisions the older adult wishes to make on their own, and disturbing others in needless ways based on the older adult's data.[37]

Today it is inexpensive and relatively straightforward from a technical standpoint for most people to install a "spy cam" or "nanny cam." Some older adults choose to wear digital alerts around their neck. Other possible tools can include sensors for detecting movement in the home, change in health status, remote monitoring for climate controls and power outlets, as well as voice-activated screens and speakers.[38] With the help of such systems, a caregiver can see if an older loved one has fallen. They can even get an alert if the older adult has left the stove on.[39] It is a reasonable question as to how "smart" you want an older adult's home to become. There are clear privacy implications to all such technologies, and older adults should understand them before deciding on their uses. Regardless of where one lands on that question though, it is possible to design all these tools in ways that ensure that older adults are more comfortable adopting them, especially if they have a say in controlling the use of their own data.[40]

We return to the power of stories to inform people, including older adults, about how technology works. The computing giant Apple produced a narrative called "A Day in the Life of Your Data: A Father-Daughter Day at the Playground." This short story helps show how many times your data are tracked and across how many applications used to plan and experience this trip to the playground.[41] Given how many older adults might be involved in taking children in their lives to the playground, this particular narrative is believable. Of course, this short document ends up telling the reader about Apple's tools that can stop the tracking of data within applications and so forth—it is ultimately marketing material for its products. But the story is nonetheless illuminating and may help older adults decide which types of applications they are comfortable using in the context of a simple trip to the park or out for ice cream with a young friend or grandchild. Of the many corporate educational products on privacy we reviewed, this short story from Apple might well capture the imagination of older adults given the near-universal appeal of taking a young person in their life out for a treat.

The role that governments could and should play as a matter of policy is so clear in the domain of data privacy that we feel it necessary to call out its importance here.

Even though the job of making policy should be squarely on the shoulders of policymakers themselves, we—our core audience of people who care about older adults—have a role to play in making the case for better policy in support of older adults and their ability to use new technologies. We can and should advocate for support locally for our libraries and community centers, so staff of these vital institutions can get the training and funding they need to offer programming to older adults on key technology issues such as privacy and security. We can and should press companies to do a better job of disclosing what they are collecting, what they are doing with what they collect, and what options older adults have to move their data when they want to—and to change their mind at any time about the choices they may have made at another time.

We can and should advocate for federal governments to make sound rules to allow greater control of older adults' personal data, to put more power in the hands of older consumers, and to require companies to enable older adults to move their data from one service to another without having to go back to graduate school in computing to make it happen. A few regions and countries around the world, mostly in Europe, have taken data privacy seriously, but most have not—and that includes the United States. While some places are much more forward-looking than others, there is a lot of room for improvement when it comes to data privacy policies in virtually every jurisdiction in the world. (By the way, these rules would be good to apply to all people and all data; it's just that we are writing this book about older adults.[42])

Technology is neither inherently good nor inherently evil. The technology itself does not make decisions to collect data, share those data without permission, or otherwise harm older adults. The technology itself does not provide health benefits or enable an older person to remain longer in their home. We do, people do. People design, code, and operate technological systems. People answer pleas for help that emanate from a lanyard around the neck of an older person living on their own in a rural community. In most cases, people still take a look at vitals that look a little off when they are transmitted by a tracking device touching the skin of an older adult half a world away. People send ambulances to check in on an older adult who might be having a cardiac arrest.

Our aim as authors is to help enable successful aging in a digital era for as many people as possible, with equity in mind and a desire for solutions that help everyone without regard to their ability to pay for expensive help. Our goal as a society should be to ensure that older adults—as with all people—are able to take advantage of the ways in which technology enhances life while not suffering too many of the negative consequences of their uses.

Privacy is one of the places where trade-offs touch ground. For older adults, privacy is a concern—but it is not the principal concern for most. Older adults want to be able to be in control of their data, to understand what is going on, and to make good choices about how much control they will give away relative to their private data.

For caregivers and loved ones, that means a combination of education and support in terms of analyzing the inevitable trade-offs. For Helen, it may well be worth giving up a great deal of her data privacy to be able to remain in her home. For many others, going on Facebook may not be worth the lack of control over their personal information. These trade-offs are a daily matter to navigate in a digital age. Older adults should have just as many good tools to make these decisions as everyone else. It will take work, but it is certainly possible. We should be able to make that happen in any and every community.

To be educated and be in control, older adults need information. While support networks can offer this, so can many resources online. A big challenge of the internet, however, is that much information on it is wrong. So how can we make sure that older adults are not being derailed by misinformation about how to stay safe online as they navigate the vast digital landscape? This requires digging into what we know about older adults and misinformation, the topic of the next chapter.

6

Misinformation

WHY DO SKEPTICS SPREAD FAKE NEWS?

Maria, close to ninety, lived in a housing complex with a few dozen other older adults. She was the informal "mayor" of the place—she had lived there a long time, maintained great mental and physical health, honed a keen wit and sense of humor, and seemed to know everyone. When anyone had a problem or needed something, they would come to Maria. She was so capable—she was trusted and she always got the job done. "Just ask Maria," was the mantra among her friends.

The forty or so residents of the housing complex, located in the northeast of the United States, were reasonably social. On a nice evening they might sit outside the tidy, well-maintained three-story building. The group played bingo and traded stories back and forth over regular social gatherings of various kinds. They shared rides to the grocery store, pharmacy, and the nearby senior center.

As the COVID-19 virus pummeled the globe in spring 2020, Maria became increasingly concerned about staying safe and healthy. As she aged, she had stayed focused on her good health. She had kept herself in fantastic shape, taking up yoga late in life and consistently walking outdoors when the weather permitted. She maintained a sound diet, visited doctors regularly, and took all prescribed medications. The results were excellent; she was the picture of health for a woman of her age. But COVID-19 posed a nasty new threat. No one would be immune at a housing complex for older adults if COVID-19 got a foothold.

Maria consumed news consistently. She got the latest information and followed what she believed to be sound. She would trust the local news providers as well as her doctors, friends, and relatives to share the latest material with her. She owned a computer and a smartphone, although she did not use them much. She relied more on traditional sources: the library, books, television, radio, and word of mouth.

At the housing complex, virtually everyone followed Maria's lead as soon as vaccines became available for COVID-19 in early 2021. With almost no exception, Maria and her friends made their way as soon as they could to get the free vaccines. When the boosters became available, same thing: Maria and friends went promptly to get protected against the deadly, fast-spreading virus. Incredibly, very few people contracted COVID-19 in Maria's complex. And essentially everyone got the vaccine the minute it was available.

There was, however, one vaccination holdout. Juan Pablo had recently moved into the complex. An older gentleman, he was gregarious and kind. He liked to sit outside with the others and join in on the activities. One day, however, trouble struck when the topic of the vaccines arose.

Juan Pablo announced that he was not vaccinated and did not plan to be. A hush fell over the ordinarily chatty group.

Juan Pablo launched into a series of reasons for not getting vaccinated, which he had gleaned from a range of sources. He was concerned about the ulterior motives of the companies developing the vaccines, the purported lack of safety associated with the vaccines, and so forth. Juan Pablo trafficked in a wide range of conspiracy theories from the internet and elsewhere about the vaccines, none of which had a significant factual basis—certainly not enough of a basis to overcome the potential health benefits of protecting himself from the harms of the virus. Nonetheless, because Juan Pablo had heard these theories repeatedly, he had become firm in his belief in their veracity.

Maria and her group of older adult friends were having none of it. Everyone else had gotten the vaccine and boosters as soon as they had become available. In fact, some of them had traveled together to the local sports arena for the free public vaccination sessions set up with priority service for older adults in the area. Juan Pablo's theories—including all the money that companies and people were making off the vaccine—did not resonate with Maria and her neighbors.

Understandably, Maria and the other residents were now anxious about being around Juan Pablo. It was bad enough to have the virus out there in the world. Now they had to listen to what they heard as fake news about the vaccinations. And what they absolutely did not want to have happen was to get sick. Juan Pablo's refusal to get vaccinated, they feared, put everyone else at risk in the housing complex. While they did not shun their new neighbor completely—and they continued to have warm feelings toward him—they also began to keep their distance when he entered communal areas of their housing unit. A bingo game with him was just not worth the risk if he was not willing to get vaccinated, they reasoned.

The risk to Juan Pablo from the false information was real, too. He was putting himself at unnecessary health risk by failing to take the free vaccines and boosters—easily accessible via prearranged rides, nearby health facilities, and fully covered costs. But he was also running the risk of greater isolation as his community began to keep a distance. Social interactions are extremely important to well-being. Juan Pablo was missing out on easily accessible nearby social connections due to his belief in misinformation he had picked up about the virus.

Before we go further, let us settle on a definition for "misinformation," the term we use in this chapter and throughout the book. "Misinformation" refers to false or misleading information, without regard to whether the originator of the content *intended* to misinform the recipient. There is a related term, "disinformation," which refers to false or misleading information where the *intent* of the entity creating or spreading the content is to mislead. This entity can be a governmental body, a corporation, a group, or an individual. A government may want to obscure the truth about what really happened in a particular situation, a corporation may want to confuse people about certain product characteristics, a group or an individual may want to inoculate others with a particular political ideology either to create chaos for their own amusement or to manipulate a situation in a way that benefits them. So while "disinformation" specifically suggests mal intent, "misinformation" is neutral as to the reason for the incorrect content. Misinformation can occur simply because someone is not knowledgeable about the topic at hand and inadvertently spreads faulty material without purposefully trying to mislead others.

In addition to the words "misinformation" and "disinformation," other terms are often used to refer to related concepts. Terms such as "fake news" and "propaganda" tend to refer to the category of "disinformation," suggesting an intent to mislead and confuse. Political scientists have a lot to say about the nuances of these terms and their role in public communication.[1] Since the goal of this chapter is not to disentangle intent on the part of the communicator, we rely on the broader term of "misinformation," which does not presume the goal of deception on the part of the originator or spreader of the content such as a person sharing a link on social media.

There is no shortage of media headlines about how older people, like Juan Pablo, fall for misinformation:

- "Older adults more likely than young to be fooled by 'fake news,' study says" alerts HealthDay News[2]
- "Older people more likely to share fake news on Facebook, study finds" heralds the *Guardian*[3]
- "Older people spread more fake news, a deadly habit in the COVID-19 pandemic" warns the *Los Angeles Times*[4]

But is this truly the case? Scholarly research on the relationship between age and misinformation is mixed. As with other topics we discuss in this book, the reality of the situation is often more complex than what newspaper headlines convey. In short, reader beware.

Analyzing data about the 2016 US presidential elections, researchers found that older adults were much more likely to share fake news content on social media.[5] This was a widely reported study about sharing links on Facebook showing that those over sixty-five were much more likely to share content from fake news domains than the youngest group. Even compared to the next-oldest age group (those just somewhat younger than them), the oldest old still shared such links more. It is noteworthy that these analyses took into consideration other important factors that may explain sharing, such as political ideology, education level, and number of links shared on Facebook more generally. What the authors also highlight, however, is this: "the vast majority of Facebook users . . . did not share any articles from fake news domains in 2016 at all." So yes, while older adults were more likely to do so, on the whole, it was an extremely rare phenomenon.

Whether due to the considerable press coverage that such research receives or more universal stereotypes about older people's technology use, which we touch on in earlier chapters, there seems to be a general assumption that older people are more susceptible to misinformation online and otherwise. This turns out not to be true, however.

There is surprising and helpful news in the data: studies have found that older adults are in many cases *better*, not worse, at assessing whether information is reliable. Research conducted in several countries about information surrounding COVID-19 demonstrated that older adults were *less* likely to believe in misinformation—at least in the domain of public health,[6] which we will discuss in more detail in a bit.

Back to our opening story about Maria, Juan Pablo, and their friends: contrary to what you may have been led to believe, Maria and her friends who trusted the information about the COVID-19 vaccines and boosters are more the norm than Juan Pablo, for whom the misinformation was persuasive (and tempting to share). Even in the case of the widely reported study about politics and misinformation,[7] only about 11% of older adults shared misinformation, and the vast majority of that small group did so only once or twice. So the practice is far from the norm overall. That is in part why our opening story is more about Maria, who does not fall for the misinformation, than about Juan Pablo, who does. When it comes to health, older adults are often better at avoiding falling for misinformation than younger people. It would be a mistake to walk away concluding that older adults are all gullible, as so much public rhetoric seems to suggest.

The problems caused by misinformation can be negligible—

getting to a store after it has closed because Google's directory listed the wrong opening hours—to profound: the incitement to violence due to misleading information about a political event or having an allergic reaction to a food product whose ingredients were not properly disclosed. There can be serious ramifications for those who fall for misinformation at any age. Governments, companies, researchers—many people are focused on this problem in the social media era given how quickly and freely information can spread on such platforms. The solutions are hard to come by, and the impact of misinformation will continue to be significant if we do not figure out some better approaches to quality assessment of information dissemination soon.

Misinformation can have particularly harmful consequences for older adults when it comes to their health. Older adults typically face more health challenges than their younger and healthier counterparts. Health is an area ripe for the spread of misinformation using modern technologies and otherwise. People of all ages spend hours searching online for possible causes of symptoms they or their loved ones are experiencing,[8] and while older adults are less likely to do this than younger cohorts, many do inform themselves this way (see more on this in the next chapter).

As older adults peruse the internet, they encounter information of varying degrees of accuracy. This experience often falls on the benign end of the spectrum as the result may simply mention some incorrect information, which a medical professional can quickly debunk assuming the patient visits one. Of course, if online searches inspire someone to neglect consulting a health care provider for their professional assessment, then even otherwise simple instances of online falsehoods can result in serious consequences.

On the more pernicious end, consider the widespread misinformation about the effectiveness of vaccines during a pandemic. Have you heard any of the following statements anywhere?

The COVID-19 vaccines were developed too quickly; they cannot be trusted.
The vaccines contain fetal cells.
The vaccines cause infertility.
You definitely shouldn't take the vaccine if you are pregnant!
The vaccines will alter your DNA.
The vaccines use a live version of the coronavirus and will lead to your getting the virus.
The vaccines use microchips.
The vaccines cause you to become magnetic.
Vitamin C will help cure you of COVID-19.

None of these statements is true, yet each can be found repeated online. These myths are so prevalent that the World Health Organization (WHO) declared an "infodemic" in the midst of the COVID-19 global pandemic.[9] WHO started addressing these myths head on by listing them on its website, explicitly explaining why they are untrue. Let this sink in. The World Health Organization, whose primary agenda is the physical and mental health of persons worldwide, focused attention on the information arena as crucial to health in the midst of a pandemic. Rather than simply sharing information about what the virus is and ways to prevent getting COVID-19, WHO was so alarmed about widely spread misinformation that it devoted space on its website specifically to debunk myths about it.[10]

People of all ages have been at risk from misinformation about COVID-19 and other health issues since the beginning of the pandemic. A study by the Kaiser Family Foundation showed that a majority of American adults (54%) in 2021 believed one or more forms of misinformation about the COVID-19 vaccines. Eight in ten of those who said they would not get the vaccine believed in one or more of the prevalent forms of misinformation about the vaccines.[11]

Studies from around the world have mostly shown that older adults were less likely to believe COVID-related misinformation than younger people. Research on misconceptions about the virus among people in the United States, Switzerland, and Italy found this to be the case even when considering differences in people's gender, education, income, and disability.[12] Another cross-national study, this time of the United Kingdom and Brazil, also found that older adults were less likely to believe in misinformation measured as the belief that garlic could cure the respondent of the coronavirus (it cannot).[13] Yet one more comparative study—of Ireland, Mexico, Spain, the United Kingdom, and the United States—found that older adults were less susceptible overall to health misinformation such as false viral videos on YouTube than younger people in four of the five countries with Mexico being the exception.[14] The researchers did not have a clear explanation for why older adults in Mexico might have been more susceptible to false health information than their counterparts in other countries other than a footnote that indicated that the sample in Mexico skewed much younger than their samples in the four other countries.

For an older adult such as Juan Pablo, the ramifications of falling for misinformation—and in turn ignoring sound public health advice or the direct suggestions of one's doctor—can be significant for health along multiple dimensions. Research has shown clearly that misinformation beliefs can lead to a lower likelihood of abiding by recommendations to stay safe by sheltering at home during the early days of the pandemic[15] and to getting the COVID-19 vaccine.[16] In turn, Juan Pablo in particular, and his peers by way of their proximity to him, may have had an increased risk of contracting COVID-19 due to his not accepting the vaccines or booster shots.

There are further threats of misinformation for Juan Pablo and others who believe falsehoods. As we argue throughout this book and as we discuss in more detail in the next chapter on well-being, one of the risks to older adults is social isolation. To the extent that misinformation led his peers to keep their distance, Juan Pablo ran the risk not just of contracting COVID-19 but also of suffering the ill effects of potentially greater isolation as he aged.

The good news for older adults is that, in general, they hold fewer misconceptions about health than younger adults. While many of us have conditioned ourselves to worry about older adults when it comes to misinformation, older people have actually navigated this treacherous information environment more effectively than many others. That is why we spent time describing Maria as well as Juan Pablo in the opening anecdote; Maria is more the norm, across virtually all studies throughout the world, than is Juan Pablo in these respects. The information-seeking and analysis skills of many older adults have actually served them quite well despite the perils and challenges of the COVID-19 period. Some of the other findings from the data about older adults and misinformation may prove equally surprising.

As a general matter, falsehoods spread more quickly than truths in the social media arena regardless of how old users are—meaning the risks that we all face from misinformation are real and potentially consequential.[17] The stakes are higher and the challenges greater these days than in the past because of the increased access to misinformation, the speed and frequency of its spread, and the kinds of decisions that can turn on misinformation. No one in society is immune from these risks. But the way these risks play out, and who falls for the traps, may not correspond exactly to what people have come to believe.

The popular myth too often takes the form of the hapless older person falling for scams and false information. This myth about older adults predates new technology, of course. These days, we may be even more likely to believe that older people are falling for frauds and passing along false information online than we did pre-internet given that some of them have fewer experiences with new technologies and stereotypically are not that engaged on social media (recall that we debunked some of these myths in the chapter on technology adoption).

The good news: data show that this myth of the gullible older adult, mystified especially by the new technological environment, is often flat wrong. Yes, older adults are too often the target for online scammers, and yes, many older adults are at real risk of harm from this targeting due to cognitive decline and other factors (we discussed this in detail in the chapter on safety and security). But what is also true is that there are certain areas in which older adults are better able to distinguish true from false information and to avoid scams that target them. Certain abilities improve, not decrease, with age. These findings do not diminish the risks that older adults face, but they do suggest that younger people may have things to learn from older adults, just as older adults often need support in navigating increasingly complex information environments.

The lived experience and deep knowledge of older people improve their capacity to distinguish fake from true headlines.[18] Researchers have shown that experience with news and information over longer periods creates an advantage for older adults. It only stands to reason: compare someone who has been reading a newspaper, living in a small community, and voting all their life to someone who is in their late teens, has never read a newspaper regularly, and never voted. The experience of the older person often results in more context for the headlines they encounter at this later stage of life. The younger person may have a better feel for a particular technology, but knowing how to navigate a particular app does not necessarily translate into being savvy about the *content* one encounters on an app.

Together, the combination of knowledge and experiences across younger and older people might result in a relatively effective way to distinguish true from false headlines in a digital environment—one of many examples that help make the case for positive, multigenerational interaction that we call for throughout this book. Another angle on this theme about older adults and how they differ: one study showed that older adults (in this case, in the United Kingdom) were less likely to trust new media than they were to trust traditional media, which tends to have more editors and layers of review than, say, Facebook or Twitter.[19]

As a reminder, susceptibility to consumer fraud is no more common among older adults than among younger adults, and if anything the data suggest that consumer fraud may actually decrease with age, contrary to public opinion.[20] We do not explore this in detail here as we considered it in depth earlier, but it is worth a reminder since consumer fraud is a type of misinformation that experience and skepticism can counter.

While it is important to resist the myths about older adults, it is equally important to acknowledge the places where they face additional risks compared to younger people. Misinformation is a threat to everyone; there are a few areas where older adults face an even greater threat than their younger counterparts. Data suggest that two areas may be a particular concern: certain types of misinformation, for instance, with respect to politics or health; and certain types of media, for instance, photographs. In an earlier chapter, we discuss the specific issues concerning safety and security since we know that older adults are increasingly the target of certain scammers, thereby facing additional risk in that domain.

A widely reported study of older adults during the 2016 US presidential election showed that they were more likely to share fake news online than younger people.[21] This study did not fully explain the phenomenon of *why* older people might at once be more skeptical, better able to distinguish false headlines, and less susceptible to false health information—and yet more likely to share false political news. One factor that may be worth adding here to the mix: older adults may be exposed to more fake news than other people, as one study of Twitter in the 2016 US presidential election suggested.[22]

Just as older adults may be more likely to share certain types of misinformation, it may be the case that certain formats cause older adults more trouble than others. One study found that older adults were less likely to identify certain manipulations to images such as the addition of something fake.[23] One advantage that younger people may have is familiarity with how image technologies work. People who regularly use image apps to brush up their photos may be more likely to anticipate changes to pictures they view than those who are mostly used to seeing images in professional settings and are less likely to edit their own photos. Early work has also found age-based differences in the ability to recognize AI speech compared to human speech, where older adults were less likely to do so.[24]

Much remains to be done in studying whether and how people of different ages and other demographic characteristics identify manipulated images. Even less scholarship exists on how people interpret manipulated audio and video. Altered images, audio clips, or entirely fake videos are going to be increasingly common in the age of artificial intelligence where it is no longer necessary to start out with a real visual to create something entirely fake yet believable enough. Academic research on how everyday people approach such content is still in its infancy. As newer forms of media replace the simple text of a printed newspaper, the forms of misinformation—and their likelihood to hoodwink adults, older or otherwise—is likely to grow with time, requiring interventions from policies to educational and support initiatives (more on this in a bit).

We know that older adults are sometimes better and sometimes less able to distinguish misinformation from the truth. Researchers point to three major factors that help explain these differences: cognitive issues, digital (il)literacy, and social change. After exploring each of these relevant factors, we dig into a paradox that some of the data show: why might older adults at once be more skeptical of certain information and yet share misinformation to a greater degree than others, especially when it comes to political misinformation?

It is no secret that older adults are prone to suffer from cognitive decline over time. Starting with age sixty, for instance, the risk of dementia increases sharply with age. A systematic review by twelve experts of data from every WHO region showed that the prevalence of dementia globally increases dramatically after age sixty, doubling with every five years of age to more than 30% after age 85.[25]

Rates of digital literacy are also an essential factor in whether older adults are more or less likely to believe, forward, and act on misinformation. Fewer older adults have used social media for many years and are often less well trained in methods for detecting online falsehoods as a result.[26] Through their responses to surveys, older adults also have shown a misunderstanding of how algorithms populate news feeds and how sharing of information can imply endorsement, although, again, younger cohorts have their fair share of confusion in these domains.[27] This is also where that insight about the digital images comes back into play. People exhibit a bias to accept images as real, so many manipulated photos can go undetected according to researchers.[28] Lower rates of digital literacy also mean that older adults may follow questionable accounts that may be not people but bots, which are automated programs that periodically post content pretending to be written by a human. Because they seem like real accounts, they raise the potential of their followers to be exposed to fake news.[29] This dynamic is especially important with respect to matters of safety and security as we discuss earlier, since older adults may be less able to distinguish a phishing message (one that pretends to be from a trusted source, but whose purpose is to steal information) from a legitimate email, for instance.

Finally, changes in the social landscape are a key factor when it comes to the relative vulnerability of older adults to misinformation. Older adults have a changing focus and changing needs as compared to those who are younger. It is possible, for instance, that older people care more about connecting to others than they do about the veracity of the information they are sharing. They also tend to trust other people more than younger people do. Research shows that interpersonal trust increases with age.[30] That is for better and for worse when it comes to the spread of misinformation.

This aspect of the changing social context for older adults is one possible interpretation of why older people might share more false information in the context of a hotly contested election cycle such as 2016: they want more "likes" and attention online, just as other people do, and they are willing to run the risk of sharing false information to connect with others in this way. This is one way to explain a puzzling set of findings: that older adults may be better able to sort true from false information while also being more likely to share false information. Another explanation is more nuanced.

Research has called this the "misinformation paradox."[31] For political content, we have already cited research showing that older adults are more likely to share fake news on Facebook even if the general prevalence of that activity is very low.[32] At the same time, many older adult participants in one study explained that they are very skeptical of mainstream media.[33] This results in the following paradox: why share content when you do not trust its source?

When it comes to paradoxes in people's online behaviors, researchers have pointed for years to a paradox concerning digital media use and privacy,[34] which we detail in the previous chapter. In the case of the privacy paradox (not reserved to older adults, by the way—it affects all of us), most people are (1) concerned about their privacy, but (2) nonetheless act in ways that may compromise their privacy. In the case of privacy, data show that it does not mean that people do not care about their privacy but that they have reached a certain level of cynicism and apathy with respect to their digital information.[35]

The same reasoning may well apply for older adults when it comes to the misinformation paradox.[36] Older adults may well have reason to believe that the information is—or at least could be—false but at the same time may have reached a level of cynicism or apathy with respect to political change. For some adults, they have spent their entire lives engaging with news and information and exercising their civic duty as voters and volunteers, only to see little meaningful change. Consider someone who may have fought for voting rights for people of color in the 1960s in the United States, only to see efforts toward voter suppression persist into the 2020s and, in some respects, only get more pernicious with time. Or perhaps the older adult has lived in a rural part of the country, heard promises for a generation from politicians during election season, and rather than see improvement, they have instead experienced relative economic decline over the generations.

Or perhaps the social context is most important: the perceived social benefits of sharing information online—even if later proved false—is worth the attention and initial connection despite the corresponding risk of being wrong. There is research to back up this final idea: older adults often prioritize their interpersonal connections over seeking new information and ascertaining its accuracy.[37]

What this means is that older adults are sharing misinformation not necessarily because they believe it to be true but for the shock value, as a way to signal to their friends how crazy some things are in the world. What may be different in older adults' sharing behavior compared to younger adults' is that older adults may share without adding their own comment to clarify their outrage and rather let the content speak for itself. (While Facebook would have the answer to whether there is such an age difference in attaching comments to shared links, we have not been able to identify any publicly available studies to confirm or challenge this idea.) If content is shared without commentary, then that makes it harder for the audience to know whether the user shared it because they believe the content is true or because they are dismayed by it and simply want to alert their friends to it or discuss the absurdity of it.

Misinformation is a huge challenge for every society in the digital age, and it concerns the population at large, not just its oldest members. As with other problems of this scale in technology, there is no easy, single solution. Addressing it requires action at different societal levels, from policymakers to technology companies to educators and caretakers.

We make the case throughout this book that many of the challenges facing older adults who are pursuing successful aging in a digital era are the same challenges facing people of other ages. While there are some differences when it comes to certain older adults—those who have not worked or have lived their entire lives without an online component, for instance—there are probably more similarities than differences overall. That is especially true in this domain: in the case of mis- and disinformation, the similarities outweigh the differences between older adults and younger people. The problems are more or less the same; the challenges of solving those problems are roughly as complex, with a few extra wrinkles in some cases.

Experts agree: no one person, company, or political actor can defeat the scourge of misinformation. It has to be a coordinated effort. Social media platforms and other technology companies need to share more data than they have in the past—as an industry, they have been terrible about working with researchers in a productive, open way. These companies also control the algorithms that too often recommend misinformation to users on their platforms—even users who have never trafficked in, say, conspiracy theories. Companies can set policies for banning accounts that promote and share misinformation that causes harm. Governments can and should set stricter rules around misinformation and should create an environment in which companies are encouraged to do the right thing rather than to fan the flames of misinformation for profit. Schools can and should do more, through ordinary classes and special sessions, to build skills among young people to ensure that they are able to discern true from inaccurate information. Libraries can offer more support for news literacy and help older adults, among others, sift through the complicated information environment with greater skill. Topics to cover include the difference between factual reporting versus opinion pieces, how to determine who is behind a website that is sharing information, recognizing that images and videos may be manipulated, and strategies to discern their credibility.

Older adults and their caregivers have a role to play in combating the blight of misinformation as well. There are a few promising strategies worth considering. The simplest advice for older adults when it comes to information online is "check before you act or share." Alternatively, if sharing for shock value, then it is best to add commentary to that effect rather than having the link speak for itself so that the audience knows the act of sharing is not an act of endorsement.

If something seems controversial, it is helpful to verify whether it is legitimate. An online search can offer a few different points of view. For specific claims, the website Snopes .com can provide very helpful information. If there are possible other arguments, then there is reason to check multiple sources or ask someone else for assistance. In many places, a librarian will answer a call or email to help assess information quality.

It is important not to assume that a quick Google search will have the needed answers, however. Communication scholar

Francesca Tripodi has shown in her research that when people rely on ideologically tainted keywords for their searches, the search results will simply reinforce their existing beliefs.[38] Accordingly, one of the important points to convey to users is that they should try different terms and keywords when looking into the legitimacy of information.

Another promising strategy to handle misinformation is the concept of "prebunking" myths. Prebunking refers to the idea of finding ways to inoculate users against the potentially harmful effects of mis- and disinformation *before* such exposure happens. Think about the idea of "debunking" myths after the fact and then just change the timing of the intervention. When a major event is underway—such as a pandemic twinned with an infodemic or a hotly contested election cycle—it may make sense for older adults and their caregivers, friends, and relatives to plan ahead.

Those in the workplace will recognize this approach: it is a form of employee training, frequently used to prevent staff from falling for phishing attacks, for instance, as we discuss in the chapter on safety and security. Some workplaces require periodic training that has a "prebunking" quality, which shows the attacks a staff person can expect to encounter. These training techniques enable the technology user to recognize the hallmarks of mis- or disinformation, or a phishing attack for that matter, and therefore to develop a healthy skepticism for when they finally encounter it.

Prebunking can be as simple as raising a key topic and getting out the positive and accurate messages before older adults are likely to encounter the misinformation online. That is always a key strategy: ensuring that audiences hear the most important, accurate information before the inaccurate information. And repetition matters. Hearing the same thing that is true multiple times before encountering the false information once can make a big difference.

If you do not feel confident of your own skills in helping with prebunking, there are online services that can assist. Games can be useful in this regard. Returning to our opening example of COVID-19 mis- and disinformation, researchers have developed a game called Go Viral! to help people recognize dangerous posts about health.[39] A similar game, Harmony Square, tackles political and electoral misinformation, an area some researchers think is a particular weak point for older adults. Bad News is a third game that helps people recognize mis- and disinformation in the news media and social media.[40]

These games are free and easy to play for those who are familiar with the technology and the genre of social media. Go Viral! takes only about ten or fifteen minutes and teaches the player to understand how those who wish to spread false information go about their task. The player is put into the role of the information spreader and prompted to consider what to post on social media, what to repost from others, and how to respond to fact-checking about their postings. These games are very new; the results are probably too fresh to draw any long-term conclusions about their efficacy, say, for older adults, who may not have as much experience with games of this sort and some of the language involving social media. But the concept of prebunking surely has promise in thinking about how to ensure that serious scientific research dominates over mis- and disinformation in the health arena, for instance.

There are promising results from a training intervention teaching digital literacy skills. An experimental study of several hundred older adults during the 2020 US presidential elections tested whether a self-directed online course could improve older adults' ability to detect false news headlines.[41] Participants were asked to assess whether news headlines were true or false at two points in time. A portion of these participants completed a training between the two tests, while another group did not. People who got the training did much better with detecting false headlines the second time around than people in the control group who received no training in between the two tests.

The online course included several pointers on how to do better in assessing news headlines. One of these is the act of opening different browser tabs in which to conduct research to see what other sources may say about the headline's claims. Another strategy is to show so-called click restraint and not necessarily click on the first search result. The course also taught users to conduct reverse image searches and identify Wikipedia page features (such as viewing edit histories and references). These proved to be helpful in improving users' ability to discern false headlines.

In the end, older adults and those who care for them will have to fend for themselves to some degree. There is no retraining program that is going to reach all older people to understand how social media algorithms really work (for one thing, no one out there actually knows this for all platforms); there is no government regulation that will sweep through and end the problem of misinformation once and for all (the media ecology is too complex for a policy to address all possible sources of misinformation). Policy approaches are necessary during this infodemic, but they are not sufficient. That means that the decision-making is going to happen largely in small groups, between a caregiver and an older adult, or one by one by the older adults themselves on their own. Our opening anecdote makes this point: Maria, Juan Pablo, and their friends are often sifting through information themselves and then discussing it at a small-group level. In these settings, decisions are made and will continue to be made about whether to believe, share, and act on information.

The data show that older people tend to be more skeptical than young people. This skepticism can be used to their advantage. If something does not make sense, then maybe it is in fact nonsensical. Misinformation has been around since time immemorial, but access to such a wide array of misinformation has never been as great as it is today through a Google search or a Facebook newsfeed. If older adults want to share outrageous content they encounter, they should do so with commentary that clarifies their astonishment and disbelief rather than letting the information speak for itself, which might suggest that the poster takes it at face value.

Against this backdrop and growing challenge, older adults need to lean into their support network and be encouraged to take advantage of those who can offer assistance in sorting true from false in the digital environment—ideally before they share the misinformation further or, even worse, act on the misinformation with potential negative consequences.

People of all ages have trouble knowing what is true versus patently false online. Older people seem to have greater levels of skepticism that helps them not fall for false health claims, but they do seem to spread content from fake news sources more readily than others (although, to be sure, at very low numbers). No matter what, older adults need to develop and refine their skills in terms of assessing information to be able to stay engaged in their communities, participate in civic life, maintain good social relationships, and ensure sound health outcomes. Successful aging turns in no small part on being able to handle digital information. This challenge is especially acute since older adults have not grown up with some of these technologies. But their lifelong skills and experiences can be put to work to help—and in turn, when shared with their younger friends and relatives, to help everybody make better decisions.

One of the key domains where information quality is of paramount importance is health. Whether digital technologies are ultimately good or bad for well-being is the subject of the next chapter.

7

Well-Being

DOES TECH INCREASE LONELINESS?

Antonia is in her late sixties, and soon after a recent divorce, a damaging heart condition began to take its toll. As she has struggled with her deteriorating health, Antonia's connections to other people have become more and more important to her. No longer able to live in her home by herself, Antonia moved into a care facility closer to her extended family. The facility promised to provide the care she would need if her health continued to decline. Removed from her home, Antonia now fears isolation as much as she fears the disease that threatens her physical health. She worries she will become increasingly isolated from friends and community members.

In the months that have followed moving to the care facility, social media has become a crucial lifeline for Antonia. She had never worked in an office, and neither had access to nor received training to use modern technology. Nonetheless, she is fortunate in several regards: though living in a rural area, broadband is good and accessible at her location, and her care home has staff that can assist with the basics of technology. Her modest smartphone and data plan, as well as her old laptop, have become crucial lifelines for her.

Antonia favors Facebook. She has a few hundred friends on

the platform, and she checks her news feed many times a day. She posts frequently herself, comments and likes the pages of others, and explores a wide range of topics on the platform. She has figured out how to vary the types of "likes" she shares—using hearts and tears as well as the basic "like" reactions. With constant use, she has become quite skilled with Facebook.

Antonia always loved the outdoors and the beauty of nature with the changing seasons in her northeastern town. As her health started to fail her, she could no longer walk outside as easily despite her fiercely defiant spirit. If the weather is too hot or too cold, or if she does not feel well—which has been increasingly the case—Antonia can access images, videos, and stories of the outdoors.

Antonia's Facebook activity reflects this engagement with nature online: her posts often take the form of a beautiful image she saw elsewhere, rounded out with a bit of text from herself or someone else. The words are often a daily affirmation, an acknowledgment of a big idea or theme beyond her reach, or a kind wish for someone else. One October day, she reposted a picture shared by someone else of a string of white lights running down stone steps next to a row of big maple trees. The image had the following quote superimposed on it: "Some souls leave behind a trail of light that is never forgotten." As she shared this image, she added: "And some have colors!" Her big heart would shine through these pictures and the accompanying text.

For Antonia, Facebook offers a way to connect to the people and things she can no longer enjoy directly. There is a wistfulness—sometimes even a crankiness—that one can pick up in some of her posts and comments. But the posts also convey a sense of joy, a gratitude that the technology allows her to stay connected to people and places that have become dramatically harder to reach physically. A November post featured beautiful pink flowers in a pot with the following note from her: "One of the last of the season!" Thirty people in Antonia's Facebook network added a mix of thumbs-up and heart reactions to this post with some also commenting on it, feedback from a considerably larger number of people she cares about than would have been possible to get in her physical presence.

The Facebook connections that Antonia has rekindled have also led to phone calls and the occasional Zoom meetings. Antonia is not especially facile with Zoom and sometimes must ask for help from others at the care home; they are happy to help her connect onto a group Zoom chat with friends and family, and Antonia rarely ends up missing one. Accordingly, she was able to connect with her godson one holiday, for which she expressed her excitement in a follow-up email: "It was great to Zoom with you on Thanksgiving! Love, Antonia."

During the winter holiday season, she also did not feel up to seeing people or traveling to a nearby relative's home, but Zoom enabled her to open presents in front of the gift giver as well as to see her nieces and nephews, grand-nieces and grand-nephews open the gifts she had given them. While experiences over Zoom are not as deep as in-person experiences would be, they are nonetheless much better than having no connection to people while unable to travel. Digital media helped increase Antonia's holiday happiness over what would have been possible without the technological intervention.

Antonia's late sixties are not all that her fifties had been in terms of social and emotional well-being. But social media and related technologies, including Facebook and Zoom, play an important role in allowing her to stay connected, stave off loneliness, and focus on her social and emotional well-being.

People often have different things in mind when they talk about doing well or feeling healthy. Scholars, too, break down these ideas, and we work from established research taxonomies that we adapt for our purposes here in this book. We have looked in depth at four aspects of health and well-being: emotional, psychological, social, and physical.[1]

It might be helpful to break down some of these health and well-being factors based on how they manifest in everyday life. When we talk about emotional well-being, we see it manifest in terms of happiness and life satisfaction. Psychological well-being might be the trickiest to understand, but consider dynamics such as self-acceptance, personal growth over time, and finding meaning in everyday things. In terms of social well-being, consider a sense of belonging and integration into a broader community and in self-actualization—the ability to be a part of improving societal conditions for oneself and others. Finally, physical well-being manifests itself as general levels of physical ability to get around, fitness, and freedom from pain. Based on the research done to date, we can say the most about how internet use relates to emotional and social well-being, and a bit about physical well-being.

Before we can examine how technology use relates to people's well-being, we need to assess how people are doing in general. We measured seven aspects of well-being in our survey of 4,000 American older adults ages sixty to ninety-four to establish how study participants are doing. (The appendix explains how each of these was measured.) Life satisfaction, eudaemonic well-being (i.e., finding meaning in everyday things), and social connectedness all skew left, which means that more people are at the positive end of the measures. Anxiety and depression skew right, which means that the majority of people exhibit low levels of both. Regarding health, two in five rate their health very good or excellent, a bit over a third good, and the rest fair or poor. We also asked whether survey participants were, in general, very happy, fairly happy, or not too happy. The majority (57%) came in at the middle, but considerably more (31%) reported being very happy as compared to not too happy (12%).

Not surprisingly, life satisfaction correlates positively with health, happiness, finding meaning in everyday things, and social connectedness, while it correlates negatively with anxiety and depression. It is especially important to highlight that social connectedness is more important for life satisfaction than health (a statistically significant difference in correlation), which bodes well for the connecting potential of digital technologies. Indeed, we are seeing increasing public attention being paid to the importance of avoiding isolation, clearly with good reason.

In spring 2023, US Surgeon General Dr. Vivek Murthy released an advisory titled "Our Epidemic of Loneliness and Isolation" with a framework for a "National Strategy to Advance Social Connection." The advisory points out that a significant portion of the population—including both younger and older adults—suffer from social isolation and loneliness. Beyond the obvious emotional and psychological problems of people feeling like they lack meaningful connections or a sense of belonging, there are also physical costs to such challenges. Social isolation has been linked to heart disease and stroke, type 2 diabetes, addiction, dementia, and earlier death.[2]

So how might digital technologies help older adults fight social isolation? Burt Minow (1922–1996) was born with partial hearing loss and facial paralysis, a condition the family later learned was Moebius syndrome.[3] In those days, such people were not known to live past their thirties, and the parents of such children were often told to put them in institutions. Instead, Burt's family embraced him and supported his growth. Outside of the family, life was isolating though. Thankfully, within Burt's lifetime, the internet came along to offer a reprieve. Burt's niece, Nell, realized that her uncle could form connections online where people would not be turned off by his disability. At first, Burt connected with others who shared his hobbies. As the internet matured though and more people flocked to it to find kindred spirits, Nell found a specific community of people with Moebius syndrome. For each family in that community, it was a lifeline to know that others like them existed and to learn that children with these conditions could grow up to live fulfilling lives into older ages, something that had not at all been apparent to members of the community otherwise given the negative responses parents often received when their child was diagnosed with the syndrome.

One of us (Eszter) suffers from a little-known disorder called misophonia or "selective sound sensitivity syndrome" and will tell this story in first person. Misophonia is a condition that had no name until 2001 and is not at all widely recognized (frustrating since I have memories of related experiences dating back to the early 1990s). Certain sounds—regardless of their volume—trigger fight-or-flight-level rage responses. The levels can range widely and can make run-of-the-mill basic everyday situations intolerable (fortunately, my condition is nowhere near the extreme levels some people experience, which sometimes can get so bad as to lead to self-harm). For years, people suffering from misophonia had been isolated and often told by loved ones, teachers, and coworkers that they just needed to deal with it as if the response was something rational and easily adjusted. Through occasional searching online over the years, I eventually stumbled on the term "misophonia" and then was able to look for online communities focused on it. Finding a Facebook group geared toward sharing coping mechanisms and other solutions was a tremendous relief. Although I have never been particularly active in that community, knowing that this is a true condition that other people also experience has made me feel less isolated, and seeing occasional coping tips has been helpful for managing the condition.

Nowadays, no matter how niche an ailment or an interest (more on this in the next chapter on learning), chances are that an online community (or five) exists for like-minded others to discuss it, share experiences, exchange tips, and generally connect about it. We note again the importance of awareness and confidence in using digital technologies (Eszter knew to look for a misophonia group) and a support network (Burt's niece searched for resources for him) for finding such convivial groups. If an older adult does not know of such a community or even that they can look for one, those in their support networks can help them find the relevant resources.

Older adults as a group are at risk of poor health, both mentally and physically. Particular concerns include loneliness, depression, and anxiety.[4] Of course, the same could be said of anyone who has gone through major life transitions such as the loss of a job. But when combined with social isolation, failing physical health, and the loss of friends and family to illness and death, the risks that older adults face take on dimensions that are different from, say, those in mid-adolescence or middle age. Many older adults no longer have parents and other older relatives, and they are also more likely to lose friends to death. As their physical health deteriorates, travel can pose new challenges. Often out of the workforce, they no longer have regular daily interactions with colleagues. This mix of circumstances in older age poses a greater risk of social isolation than most other age cohorts face.

The Harvard Study of Adult Development is a unique project that has been tracking the health and well-being of three generations of people for over eighty-five years. When looking at what matters for people's health and happiness most consistently, the study has found it to be good relationships through close personal connections.[5] Given how life changes in older age shrink people's personal networks, it is important to consider how connections through new technologies can help fill those voids.

The use of technology can matter a great deal to the health and well-being of those who use it, whether the technology user is an older adult or not. What is most important is not the amount of time spent online but the purpose and nature of the technology use.[6] These findings are similar to findings about other age groups and their technology usage.

Some of our work has involved extensive study of young people and their use of technology. Over several decades, we have conducted and read study after study that shows that the driving factor is the type, nature, and quality of the use of technology rather than rough factors such as the number of hours on a particular social network. And rarely do the data say something simple: it is almost always nuanced, complex, and colored by the personal experience of the individual.[7] That makes it hard to generalize from one type of online activity to another when it comes to the relationship between technology usage and health and well-being outcomes.

Here is an example of findings from a well-researched paper about older adults:[8]

> In general, spending more time on the internet was associated with higher levels of social loneliness. However, time spent online to communicate with others was linked to reduced levels of loneliness. . . . Furthermore, research highlights that online engagement for social purposes is particularly helpful to reduce loneliness.

Read those lines with care and you will see that there is insight to be found in the data about older adults and use of new technologies, but the findings can seem contradictory. They require you to think hard about what you are hearing and reading. They call on you to examine and compare methodologies used by researchers, which are not universally strong—or even always sound for the task at hand. And this exercise of looking deep into the research about older adults proves that there is no easy "one size fits all" advice that will work for everyone.[9] We will break down some of the research in more detail given just how complex the issues are.

Many studies in the social sciences look at data captured at one point in time. For example, a study might measure older adults' level of loneliness while also measuring whether they use social media. Were such a study to find that people who are on social media are lonelier, it would not allow us to answer the question: does social media use result in loneliness? The reason is that we cannot deduce anything about the direction of causality from such cross-sectional data (i.e., data collected at just one point in time without any experimental manipulation). This finding may mean that lonelier people are more likely to flock to social media, or the finding may reflect that spending time on social media results in loneliness, or a combination of the two. But we cannot say for sure which of those it is. More advanced careful studies are needed to address how time spent on social media may then link to feelings of social isolation as opposed to being a precondition to turning to it in the first place.

One elaborate survey study followed adults aged fifty-five to seventy over several years, collecting information about their demographic backgrounds, their internet uses, their loneliness levels, and their well-being at three points in time.[10] The project collected data about people's various online behaviors:

social activities such as connecting with friends and family, making new connections, and sharing photographs; informational uses such as reading the news, seeking health information, searching for entertainment; and so-called instrumental uses such as business activities, work tasks, and shopping. The study found that social uses resulted in a decrease in loneliness, whereas informational and instrumental uses did not. And not surprisingly, higher feelings of loneliness predicted a decline in well-being. But it was only social uses that reduced loneliness thereby avoiding a decline in well-being. How do we reconcile this with other studies that linked general social media use to loneliness? The devil is in the methodological details of the studies. The one we just described collected information about different ways people may use social media, whereas other studies often just look at *whether* people use social media without regard to *how*. How people use such platforms is crucial for whether that behavior has implications for their well-being.

If you found out that an older, recently widowed relative was spending a lot of time on Facebook while confined to their home with a debilitating illness, would you feel worried about them or would you encourage more time spent on social media? What can research tell us about whether you should be concerned?

Given the multiple aspects of health and well-being, and the wide variety of ways in which older adults use new technologies, there are countless approaches to analyzing this topic. There are two factors that the research suggests are strongly correlated: social isolation can result in anxiety and depression in older adults. Research shows that social connectedness, especially when it concerns sustained connections over time with the same people, in turn, can be a key factor in improving a person's emotional and social health.[11]

The strongest conclusion to come out of years of study is that positive, active online social engagement can be helpful to older adults' mental health. We adopt the idea of "online social engagement" as the key here, encompassing multiple types of online participation such as exchanges in online forums, structured discussions where people are asking and answering questions, or looking at photos or videos of family or friends.[12] These activities can be on social media platforms such as Facebook and Instagram, on sites that allow for the sharing of photos such as Shutterfly, on mailing lists or forums, or one-on-one over smartphones on FaceTime, WhatsApp, or Zoom.[13] One important point here is that it is rarely the platform that matters but what people do on it that makes a difference. Aimlessly doomscrolling is qualitatively different from engaging with the content shared by loved ones and sharing such content in return. One study found that it even makes a difference whether one is looking at the photographs of family and friends (good) as compared to images of others (not so good).[14]

Positive online social engagement can help older adults affirm and express their identity and share their emotions in helpful ways. Recall the story from the start of this chapter where we discuss the important connections Antonia was able to make even when confined to her home. These social interactions with others, in turn, can counteract loneliness and help stave off depression and anxiety. One study found that older adults have fewer connections on Facebook than younger people, but a higher portion of older adults' friends on the platform are "actual friends."[15] The study also showed that this higher ratio of actual friends as well as more active engagement with one's friends on the platform linked to higher perceived social benefits measured as feeling like "you get support from friends on social media."[16]

Social media often get a bad reputation in the media for being the cause of many problems, but nuanced investigation suggests a more complex picture. When we asked 2,000 older adults whether they have various negative (such as overwhelm, loneliness) and positive (such as connectedness and joy) feelings while using social media, many more reported positive feelings than negative ones. In fact, positive feelings were 2.5 times more likely than negative ones. The least reported positive feeling, gratitude, was still much more common at 25% than the most reported negative feeling, anxiety, at 14%. Perhaps due to the types of connections older adults cultivate on social media, their experiences tend to skew positive.

If we were forced to simplify the research as far as we can, we might say: meaningful social engagement online can—but does not necessarily—lead to better health and well-being for older adults.[17] So when it comes to that recently widowed older relative, if they are engaging actively and positively with people in their networks, then you likely have nothing to worry about. Indeed, their online activities may be helping them cope with their isolation.

In addition to well-known social media applications like Facebook, specialty services are cropping up that can provide structured, potentially very positive social interactions. One such service is called Papa. Through this service, older people and younger people are matched to develop genuine connections. This service is structured to provide help to older adults in need through "papa pals" to make meaningful social links and help with chores, brought together through technology, but then also connecting in person. The service started with its founder needing to help his grandfather but being unable

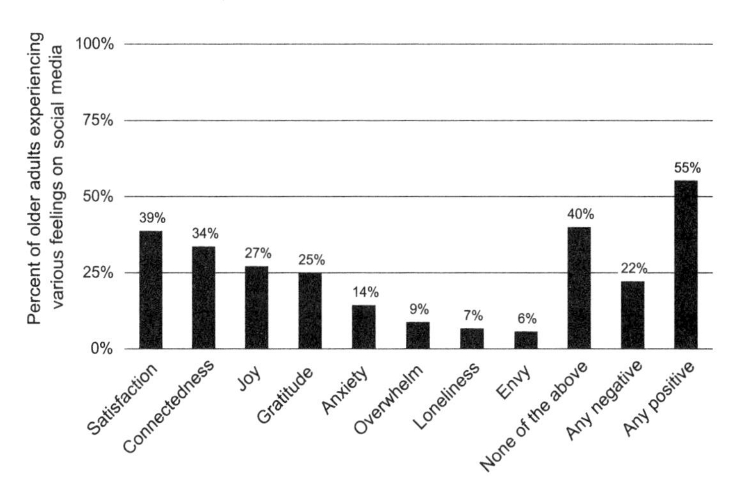

Percent of 2,000 older adults experiencing various feelings while using social media (among social media users) (2023).

to do so personally at all needed times. The app lets interested younger people set the times when they are available and find matches of "papas" (and older women) in need of assistance based on location.

Another social application called Eldera also connects people across generations. As reported in the media, the Eldera app brought together an eleven-year-old boy, Hyunseung Lee, living in Seoul, Korea, and a University of Illinois–Chicago professor, S. Jay Olshansky, to discuss their interests in science.[18] The app gave the mentor and the parents of the young mentee a Zoom link on which they could connect to share their interests. The biweekly interaction between Lee and Olshansky led in part to the young member of the pair, Lee, publishing the book *Beautiful Theorems that Changed Math*, whose sale profits have benefited a pediatric cancer organization. The founder of the Eldera app, Dana Griffin, was inspired to start such a service thanks to the positive memories she had from being raised by her grandparents. These types of mentoring arrangements have offered mutual benefits for centuries. New technologies now make them easier and can enable connections across large distances.

There are important limitations to these findings about older adults, their use of new technologies, and the connection to health and well-being. It is only fair that we draw these potential limitations to your attention too.

A deeper dive on the connection between extensive online community engagement and anxiety demonstrates these limitations. Worldwide, there is a growing prevalence of mental health conditions. Anxiety is among the most common, affecting 265 million people.[19] Mental health conditions impact approximately 20% of those age sixty and older worldwide.[20] Social isolation, which is a particular concern for older adults, is strongly associated with mental health,[21] although the relationship between social isolation and anxiety is less clear.

A study from 2018 showed a correlation between participation in certain online communities and increased levels of anxiety in adults sixty and over.[22] The survey asked how often respondents had participated in meaningful online discussions about a list of topics such as travel, religion, food, politics, and sports. This work dug more deeply into concerns based on other work about the potential for technology-related isolation.[23]

The study found that different ways of socializing online related to greater symptoms of anxiety when controlling for sociodemographic factors, social context, health, and general internet experiences and skills.[24] What we found particularly fascinating about this research is that the degree of anxiety was higher when talking about certain topics in these online communities than about others. Meaningful discussion about aging and health related to great levels of anxiety whether that concerned retirement, caregiving, a health condition the respondent had, a health condition of a loved one, general health, or exercise. Six of the eleven "general interest" topics—arts and crafts, religion, finances, food/recipes, technology, and gardening—were also linked to symptoms of anxiety. The topics that were not related to higher levels of anxiety were travel, activism, entertainment, sports, and politics. Yes, politics.

Recall, though, what we said earlier in the chapter: it is not possible to draw conclusions about the direction of causality based on survey data collected at one point in time. In other words, the old adage: "correlation does not imply causation." It may be that people who are already anxious about a health condition go online to discuss it and that is why people who discuss such conditions exhibit higher levels of anxiety. It may also be that people who get anxious about politics decide not to seek out related discussions online or do not deem to have had meaningful discussions about the topic; those who do report such experiences are ones who do not show higher levels of anxiety. Had the research focused on incidental exposure to online content as opposed to meaningful discussions of it, it may have found that anxiety relates to seeing political discussion.

One takeaway here is that the specifics of a study matter. People who are not research experts on a topic often get their exposure to scientific findings through the media. While the press can do a lot of good in disseminating information, it is often more focused on attracting eyeballs (and advertising dollars) than getting into the nuances of a study. But complex research rarely translates well to sound bites, and so important details get lost in translation. And when it turns out that the press misreported something, getting coverage for the correction is an uphill battle.

This brings back memories of the unfortunate coverage of a study concerning Facebook use and academic outcomes in 2009.[25] Some scholars had presented a paper at a conference claiming that Facebook use was linked to lower grades among students. The press ran with these findings to ring the alarm that using Facebook *resulted in* lower academic achievement. The problem was that the paper had some major shortcomings (like conflating graduate students with undergraduate students), and there was no evidence for causality even though the press coverage had implied it.

A response paper Eszter wrote with two colleagues pointed out the issues with the conference paper and offered evidence from three data sets that Facebook use was not linked to academic achievement.[26] Even though this paper was published within a few weeks of the misleading press coverage, the follow-up paper received very little media attention. Why? Probably because there was not much to be gained from saying that there was no alarm bell to sound. Your takeaway from all this? When seeing media coverage about the implications of technology use, seek out the source to see whether the media interpretation is correct.

To be sure, using digital technologies in a social manner can have dysfunctional effects on older adults' mental health. For instance, research has found links between mental health and problematic usage patterns, negative social comparisons, or when technology use is a replacement of or a compensation for other social interactions.[27]

The research picture is also not entirely clear when it comes to the relationship between older people's use of technology and their likelihood to suffer from anxiety and depression. Some studies suggest that pragmatic, purposeful use of social media can reduce a sense of loneliness and, in turn, have a positive effect relative to mental well-being.[28] In contrast, other studies have shown a connection between extensive social media use among older adults and depression.[29] Some studies have shown that the more time a person spends on a social network site, the more likely the person will show depressive symptoms,[30] although much of this research is not about older adults per se. Again, we must remember that correlation does not imply causation. Other studies have found no relationship between extensive time spent on a social network site and depression.[31]

One potential reason for the many conflicting findings is that studies measure different technologies using a variety of methods, making their comparisons a bit like comparing apples and oranges.[32] Some talk about whether any use of social media is connected with negative experiences, while other studies delve into whether interactions on specific platforms or particular types of uses correlate with anxiety or depression.[33] Some studies look at time spent on social media; others look at whether being unfriended comes with negative implications.[34] Further complicating the matter is the quickly changing nature of the object of study. Does a study about Facebook from 2010 generalize to the situation on the platform in 2020? Given that the demographics of users changed over that decade, that the platform itself evolved considerably, and that a global pandemic occurred in the latter year, it quickly becomes understandable that finding one grand truth about how social media uses impact well-being is difficult.[35]

We must be vigilant about the details of studies when extracting conclusions from them. We must accept that it is likely unrealistic to expect a universal truth about the link between digital technology uses and well-being. This is perhaps a frustrating takeaway, but a more realistic one than implying that all of technology, over vastly differing cultures and time periods, across users of different types, will always show consistent findings.

So with that in mind, what are some examples of clear takeaways? There is some evidence that text messaging nudges can lead to engagement by older adults in physical exercise, which could in turn lead to positive physical health outcomes. One study looked at a small group of older African Americans and showed, using a randomized control trial, that there was an increase in step counts and physical activity overall.[36] A second study, involving a randomized control trial of Malaysian older adults, demonstrated that text messages could be effective nudges leading to higher physical activity rates.[37] Given that lifestyle choices such as increased physical activity and, ideally, structured exercise are proven to lead to better health outcomes for older adults, these studies show promise in terms of the ability for tech-based interventions to help lead to better physical health outcomes for older adults.

There are several reasons why a text message type of intervention may work well with older adults. First, it is a fairly simple technical intervention. Most phones, no matter how sophisticated—or not—are set up to support text messages. Most data plans support at least a small number of text messages, and even those that charge per message are relatively inexpensive. Once the system has been set up, it runs by itself. There is not much that anyone needs to do. And one of the reasons that text messaging tends to be effective is that people report feeling that it is more "personal" when someone sends a text message as compared to other forms of technologically mediated communication. The text is meant for you and you alone, urging you to do something that is in your own interest. As a result, text messages that offer "nudges" have a decent shot at being effective for older adults—in turn potentially leading to better health outcomes.[38]

One thing we know for sure, across all age groups we have studied: those who are better off tend to get better results out of using new technologies. Decades of study of various forms of the digital divide have established that social, economic, and educational backgrounds matter a great deal in terms of whether an individual benefits significantly from the positive sides of using new technologies. Without question, we now know that in most settings and most ways, people from higher socioeconomic backgrounds (higher levels of education, more income) tend to benefit more from their digital media use than less privileged people do.[39]

The data are less clear on the "harms" side of the equation, such as being the victim of identity theft, losing financial resources, experiencing overload, or facing heightened surveillance.[40] The question of whether more harms accrue from new technology usage to those who are less privileged has been studied less than the questions associated with benefits, but the few existing results suggest that those who are socioeconomically disadvantaged are more likely to have negative online experiences, which yet again makes for a worrying picture about differences in older adults' digital access and use.

It is also important to note that the implications of technology use for well-being have not been studied in anything approaching a comprehensive manner. No single study has traced the full cycle of how social status, for instance, affects the way social media are used, which then may produce well-being-related outcomes that accumulate into longer-term general well-being and, in turn, feed back into other aspects of that person's life, such as opportunities and other life chances.[41] That is a long loop and hard to study—but it may make a great deal of difference as we seek to understand deeply the relationship between older adults, their technology practices, and their health and well-being. It is as though we have a series of still images that tell part of the story but no moving image to give the full context, to show the many forms of interconnection in the story that sometimes matter the most to our understanding.

The essential research finding—that older adults can use new social technologies to reduce loneliness and increase social connections, thereby reducing their risk of anxiety and depression—suggests to us as authors and researchers that there is important work to be done to understand this area better. If we can come to understand clearly the types of deliberate interactions that are best suited to address loneliness and social isolation for older adults, we can help structure both advice and possibly new forms of social technologies to meet these needs. The new slate of mentoring applications, such as Papa and Eldera, as well as simple text-based nudging, indicate what might be possible in the future, informed by sound research.

Researchers, technologists, policymakers, and caregivers also need to focus on the other side of the equation: the potential harms to older adults from engaging in similar activities. There are hints in the research that excessive use of new technologies can lead to harms to well-being, but these dynamics are not well understood. Engaging in meaningful online discussions of some topics has been linked to higher levels of anxiety,[42] for example, or could result in more feelings of loneliness, which we have shown have repercussions for physical health. And to be clear, there is a similar dynamic with respect to understanding the harms for younger people as well: no doubt some uses of social technologies can lead to harm in teenagers, for instance, but our understanding of how that works is still hotly debated among researchers, policymakers, and technologists.[43]

Overall, these studies indicate that positive, structured social experiences online can improve older adults' sense of connection to others, reducing their loneliness and the likelihood that they will grow more anxious and depressed over time. It makes sense: older people who spend time online chatting with their family or neighbors are typically happy about that. New digital technologies can help build these social connections that can be harder over time for older adults to maintain in their physical proximity. It is possible—we think likely, in fact—that social support and technical advice designed specifically for older adults when they are using new technologies can help ensure positive outcomes from a health and well-being perspective. And applications that are grounded in good health and well-being research can nudge older adults to do positive, healthy things as well as to develop new relationships that lead to increased happiness and a sense of overall well-being. The types of learning that older adults can undertake online to support such goals are numerous and we turn to discussing these in the next chapter.

8

Learning

CAN NEW TECH TEACH NEW TRICKS?

"It's clear to me this can be an exceptionally important thing for the future," Newton Minow told us not long before his passing. "All of education is going to be transformed by technology. But I think for older people, particularly those who don't get out, this is something they can do at home. It is a great social benefit for them."

Newt Minow had plenty of standing to make such a bold claim about the importance of new technologies and how they might play out in terms of their importance for older adults and learning in the future. Newton Minow—incidentally, the brother of Burt Minow mentioned in the last chapter as having benefited from an online community of people with Moebius syndrome—played a central role in the development of new technologies over a span of more than sixty years. President John F. Kennedy appointed Minow, a lawyer and civic leader, to serve as chair of the Federal Communications Commission (FCC) in 1961. He also chaired the Board of Governors of the Public Broadcasting Service (PBS).

During his momentous term as FCC chair, Minow popularized the term "the vast wasteland" to describe television and prompted extraordinary investments in mass media in the public interest. It is no exaggeration to claim that the development and spread of the modern PBS in the United States dates to his leadership at the FCC. After his time on the FCC, Minow continued to serve the public through the board of PBS and local media organizations.

In his late nineties, Minow still contributed to the dialogue about the future of technology and media. From coauthored book chapters to public speeches, Minow was no less engaged in the significant public debates in the 2020s than he was decades earlier.

What led to our interview for this book project, however, was not his fame as a technology regulator but his recent teaching assignment. Along with his daughter Mary Minow—herself a leading advocate for public access to knowledge through technology, libraries, and the law—he led three courses, known as study groups, for other older adults offered by Northwestern University.

The system that powers these courses is called OLLI—the Osher Lifelong Learning Institute. OLLI has hundreds of chapters across the nation supported by the Bernard Osher Foundation. In Chicago, the OLLI chapter is based at Northwestern University. The chapter's tagline: "Curiosity Never Retires." And OLLI at Northwestern University had an ace in the hole: the Minows were among its star faculty members leading courses for other older adults. Northwestern published a profile of the Minow father-and-daughter team with the headline, "A Rookie at 96."[1]

The story goes back several decades. In 1986, Minow was on a fellowship at Harvard University. Each day he passed a sign for the Institute for Learning and Retirement. Inspired by the idea, he advocated for a Chicago-based program of a similar sort back home at his alma mater, Northwestern University. When we interviewed him for this book, at age ninety-seven, he was teaching in the program he helped to start. With a hint of pride, Newt Minow reported that the "Osher Foundation moved the national headquarters to Northwestern."

The OLLI program makes it possible for older adults to learn about a topic of interest to them from just about anywhere, and with a relatively small amount of technological knowledge to get started. In addition to in-person classes, OLLI courses can be accessed via Zoom. As Mary Minow told us, getting started on the Zoom classes is pretty simple: "All you really do is click the link. The only thing that's hard is if your video and your microphone are turned off."

The same goes for the teachers of the course, called coordinators in the OLLI system. In the Minow household, the two coordinators—father Newt and daughter Mary—worked as a team to set up and lead the class. From a technological standpoint, they had few hurdles. Speaking of his daughter, Newt clarified that "Mary is very advanced in technology" and taught him all he needed about how to engage with the Zoom and Canvas technology platforms that they used to teach the three courses.

For each OLLI course, the Minows could supplement their team by adding another coordinator who would focus on technological support. The job of these two to four coordinators is to frame the content, lead the discussions, and support others with respect to any challenges participants face as they navigate the technologies involved. One such coordinator was Julia Katz, a longtime family friend whose presence allowed the Minows to cultivate old social ties in addition to forging new ones. As Mary reported, due to the structure of the course, "Dad was let off the hook on being the technology troubleshooter for other people" and could focus on his area of expertise in his role as coordinator, while others could handle the technology support.

For challenges that exceed the coordinators' ability to help, the OLLI program makes available a discounted service for technology support by a company called Near Technology. Near was started by undergraduate students at Northwestern University who had the idea that technology support for older adults could be a good business. As the Minows reported to us, a small monthly fee allows access to a Near technology consultant who can help with the course or other technological needs. If a private house call is needed, the technologist would come out to visit in person for a modest additional fee.

From their course on *Fiddler on the Roof* to one about *War and Peace* as it has been adapted for film, the Minow father-and-daughter team showed the way forward for older adults who want to keep learning and keep teaching. They were able to link in volunteer guest speakers seamlessly from universities as near as the South Side of Chicago to as far as Vermont. These guest speakers could log onto Zoom for fifteen minutes or so to engage with the older adults enrolled in the course and share their expertise, then pop back out and return to the rest of their day. New digital technologies meant that the Minows could keep up their teaching no matter the context outside their home: whether during a global pandemic, a freezing Chicago day in the depths of February, or a perfect spring semester.

We would not bet against Newt Minow's powers of prediction about the importance of new technologies for lifelong learning given his track record.

Newt Minow's claims are correct: when brought together properly, older adults can achieve a range of positive outcomes when they use new technologies, as the OLLI system demonstrates.[2] These benefits include boosting confidence and a sense of self-efficacy, encouraging socialization, and promoting awareness of and interest in important subjects. Some studies have also shown an uptick in a sense of general well-being by older adults as a result of online learning experiences. The downsides, including, for instance, the potential for frustration and fears of information overload, seem to pale in comparison to these substantial benefits that older adults can gain from learning through new technologies.

Despite the strong case for focusing on older adults and the promise of their online learning, leaders in societies too often do not think enough about older adults as "learners" and fail to invest in this important work sufficiently as a result. When the topic of "learning" arises, the age group that immediately pops to mind is probably young people—schoolchildren or perhaps college students—almost certainly not older adults. This is natural: much of the time, we associate "learning" with "school" or "university." The majority of research on learning and education centers on young people, with much less academic focus on the learning practices of older adults. Yet with a growing population of older adults across the globe, informal and formal learning opportunities can offer great value.

The truth is simple: important learning takes place at every stage of life. Early childhood education sets young people on the course for life. Elementary and secondary education can prove determinative in an incredibly broad range of ways for young people. Vocational learning can help launch adolescents into lucrative and fulfilling careers, while college and university educations can help others expand their horizons and get their start in professions. Within the workplace, continuous learning and professional development programs are essential for employment success, just as adult learning programs in libraries and community centers serve as important individual and social goods.

Learning for older adults can and should take place in a

mix of settings, ranging from the formal—as in the Minow example at Northwestern University or a workshop at a library or community center—to the informal—such as trying things out on one's own device at home or with some friends. No aspect of this learning trajectory should fall by the wayside, but too often older adults are left out of the conversation about the importance of learning. New technologies make learning more accessible than ever, especially for those who cannot as easily get out to take advantage of in-person learning opportunities. Older adults ought to be viewed as active creators of digital content as well as potential teachers—not just as the passive recipients of information served up by younger people using technology.

Informal learning can be an important way that older adults stay engaged with the world. Research has found that people use different learning approaches—formal versus informal—depending on the topic. Namely, when learning about health care, physical fitness, travel, fashion, and do-it-yourself projects, one study found that people are more likely to turn to informal sources such as how-to guides, Wikipedia, YouTube, and question-and-answer forums.[3] Accordingly, it is important to make sure that older adults are aware of such resources so that they can take advantage of them for informal education.

We asked our survey participants what types of online learning resources they have *heard* of. The vast majority (96%) had heard of at least one type, which is encouraging. The most widely known resources were online video sites like YouTube and TikTok. Many older adults were also familiar with Wikipedia and online reference sites like WebMD, library or museum sites, and dictionaries. Only about half knew about online discussion communities like Facebook Groups and just

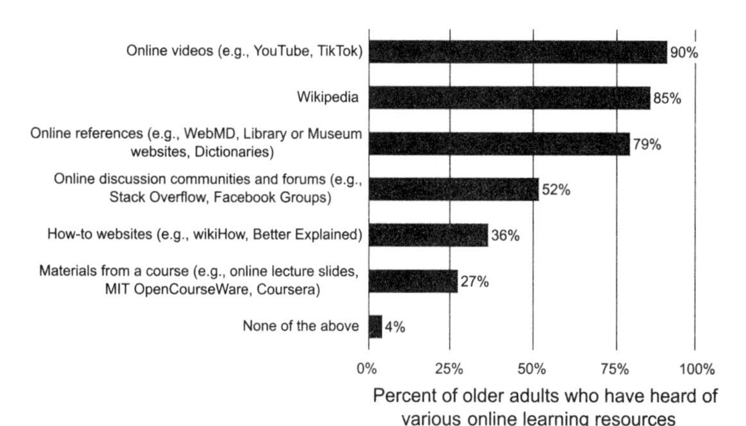

Proportion of 2,000 older adults who have *heard* of various online learning resources (2023).

over a third about how-to sites like wikiHow and Better Explained. The lowest awareness concerned more formal online resources such as lecture slides and online course platforms like MIT's OpenCourseWare and Coursera. While some of these results are encouraging overall, they also suggest that there is room for making older adults more aware of what online resources are available for learning—many of which are available for free or at extremely low cost.

Learning for older adults can be a crucial component of well-being. The importance of this population and stage of life only continues to grow as lifetimes become longer. As a United Nations study pointed out, by 2030 it is likely that the number of adults aged sixty and over will exceed the number of children aged nine and under for the first time in human history.[4] It behooves societies to offer meaningful opportunities for these millions of people.

As we have throughout this book, it is essential to note the diversity within the population of older adults. Nowhere is this variety more important than when it comes to learning about new technologies or using new technologies to learn about other things. A sixty-one-year-old still in the workplace who has worked in the technology sector for two decades, whether in Zurich, Tel Aviv, or Palo Alto, may be among the world's most tech-savvy, connected people on the planet. A ninety-one-year-old in the same community who never had an office job may feel disconnected and afraid because of the perceived need to be online and using technologies that seem foreign and frightening. Policy regimes need to take into account the heterogeneity inherent in the older adult population when it comes to new digital technologies.

As we note in the earlier chapter of this book on technology adoption, the research on older adults shows that they are often both interested in learning about new technologies and able to do so. In general, older adults learn best when they are supported at the outset by others. While it is possible for technologically savvy older adults to learn to use new technologies by themselves, studies show that these adults are in the minority.[5] There are some older adults who can figure these things out on their own and therefore do not need support from others. In turn, these older adults may be crucial community members who are able to help their peers. A major theme from the research about older adults and learning with technology: older adults can play a multitude of roles in the equation of teaching and learning, from teachers themselves to learners, often all at the same time.[6]

One study followed just over 200 older adults for three years to learn about their motivations for and benefits of learning video production through courses offered at a local community center.[7] The workshops were led by a mix of volunteer peer older adults who had studied at the center earlier and younger people. Older adults listed several motivations for embarking on such a learning path: to share important family moments, improve their technology proficiency, and feel more socially included in a society where computers have taken center stage. Interviews with the participants revealed that they found the experience positive in that it helped them gain new skills, which then made them feel more socially and digitally included. They also found it encouraging to know that they could still pick up new skills and could even help others with their learning.

Often older adults prefer using technologies that they learned earlier in life—with which they have presumably become quite comfortable—rather than learning and adopting newer technologies later in life. That does not mean they cannot learn new technologies though. It means simply that there may be a bit of an uphill struggle for those who are trying to convince older adults to try out new technologies later in life.[8]

It also may make sense to support older adults to adapt their existing technology usage to a new situation rather than trying to urge them to use a new technology to accomplish a new task. For instance, if an older adult has become used to turning to their desktop computer from a few years ago to order food online, it may not make sense to try to get them to turn in their flip phone for a smartphone, just because the latter can be loaded with food-ordering applications. For an older adult in our lives, the ability to order in on an aging desktop computer was a lifeline for years. No amount of coaxing was going to change this practice to a newer type of device unless it was truly necessary (and it wasn't). Put another way: just because a person giving support prefers a certain method to accomplish a particular goal, the older adult's preferred approach should be prioritized as it may mean the difference between benefiting from assistance that technology can offer versus not benefiting at all.

In the example involving the Minows and their lifelong learning courses, the benefits of engagement with the new technologies included but also extended beyond use of the course tools themselves. Yes, the knowledge of how to use Zoom (a videoconferencing tool) and Canvas (a course management system) might prove helpful in other settings for the older adults who used them for the OLLI course. But the people who signed up for the course were likely motivated by the topic of the discussion more than the use of the technology for its own sake. This is reminiscent of how older adults will sign up for certain apps to connect with particular people like their grandchildren. It is yet another example about how purposes often unrelated to technology drive technology adoption. Learning about the latest gadgets becomes a side benefit of the activity rather than serving as the original core mission.

In the case of these formal courses, the older adults were likely driven by their interest in stories or classics of literature. Perhaps they had never read *War and Peace* and wanted to do so. Or maybe, like the Minows themselves, they had read it a while ago and had enjoyed it very much, hoping to return to the text and the movie adaptations along with others—and now here was their chance. Or perhaps their lifelong devotion to Jewish traditions made them interested in the story-behind-the-story of *Fiddler on the Roof*. You get the point: learning to use new technologies later in life is not just about learning to use technologies. It is learning for the sake of learning, as well as ancillary benefits such as socialization with others.

In terms of less formal learning, older adults have much to

gain from new technologies as well. For instance, older adults may turn to digital tools to help with information-seeking. Even those who do not enroll in a formal course may benefit from using the internet, for instance, to learn about health. A simple example, building on the previous chapter on wellness, is learning about COVID-19 testing sites, ways to stay safe during a pandemic, availability of free tests by mail, ways to get vaccinated, and so forth.[9] While there are many sources of information on these topics, the internet can serve as a simple one-stop-shop for older adults (as it can for any person, for that matter). To take advantage of this powerful tool, of course, the older adult needs the skills, easy access to the tools, and the support network to stick with it and reap the benefits.

What exactly can one learn about online? In short, everything. In our survey, we asked participants how interested they are in various topics (unrelated to learning, just in general), and three-fourths of them indicated the following diverse subjects (in this order): food-related topics (cooking, baking, restaurants), health and fitness, politics, home improvement, entertainment, travel and geography, and science and research. All of these are themes widely represented online and available for learning. There are lots of recipe websites, cooking videos, health advice, political discussions, do-it-yourself project ideas, and so on. There is a treasure trove of video material to help those looking to learn about these topics, countless articles, and numerous discussion communities where a user can browse without getting involved or participate in exchange with other enthusiasts regularly or just on occasion. A nice instance of how digital technologies can support one's interests is Nancy Epley of Kentucky, who at age 100 was still busy researching images online to help her with her art pieces and jacket stitching, the products of which she then happily shared with her friends and family.[10]

Just to take as one example, the online forum Reddit has communities organized by topic dedicated to discussing anything from antiques to breadmaking, college sports, dating advice, embroidery, filmmaking, genealogy, hiking, insect collecting, journaling, kombucha brewing, Lego building, meditation, neuroscience, orienteering, public speaking, quilting, reading, singing, travel planning, upcycling, vinyl records, welding, yoga, and zoo visiting. As research has shown, people can build lasting connections on such forums where interactions, over time, can extend far beyond the specific topic of the group.[11]

Beyond online learning materials, digital technologies can also help with connecting people to those in their hobby communities. Antoinette is retired and lives in a small town in Switzerland. A decade ago she signed up on the Postcrossing website to send and receive postcards from around the world. It is an online service that lets people connect to other postal mail enthusiasts, not through becoming pen pals (although that is certainly an option) but through one-off postcard mailings. It serves a helpful social function by connecting people across the globe. It also allows Antoinette to learn about different cultures—not through the internet per se, but through giving her a reason to put pen to paper and affixing a postage stamp to it. It is her use of the website that gives her this window onto the wider world. Antoinette is also part of an active Facebook group where members of the Swiss Postcrossing community exchange experiences, gift each other postcards, and plan in-person meetups. Again, using Facebook is not the end goal here, but it acts as a helpful lubricant to accessing the community that shares her passion for postal mail. In the past few years, Antoinette's granddaughter has also joined in on the fun and now accompanies her to meetings. Thanks to these online possibilities, Antoinette is both engaging with and learning about the wider world while also making meaningful connections with her granddaughter. What's not to like?

A particularly relevant type of information in older age is health and health care. We asked survey participants whether they had sought a myriad of health-related content online, and the vast majority (93%) reported such experiences. Two-thirds of older adults had used the internet to seek information about a specific drug or treatment, a similar number had checked medical test results online. Quite a few (62%) had researched drug safety and side effects. Asked about eleven types of health content, the average older adult who had used the internet for any of them had used it for six types of information, showing that digital technologies play a clear role in informing older adults about health-related matters.

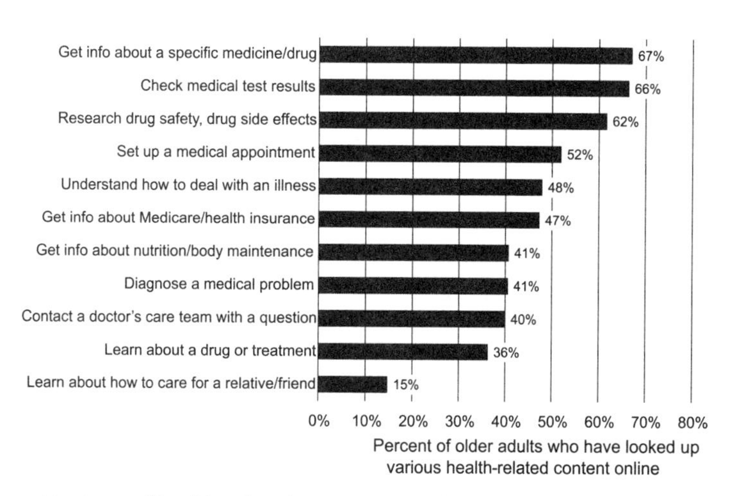

The types of health-related content 2,000 older adults have looked up online (2023).

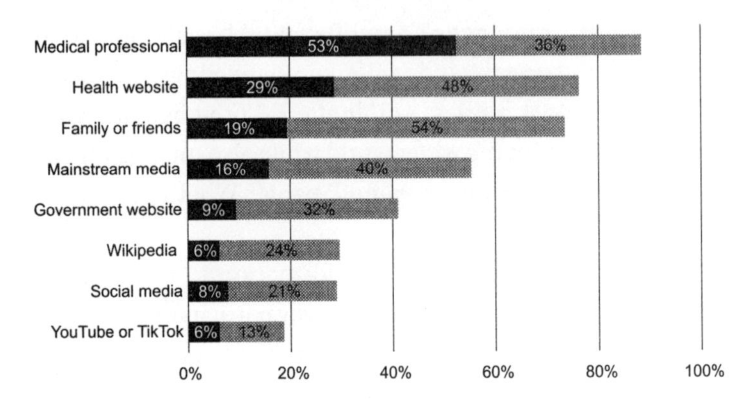

The types of sources 2,000 older adults consulted in the past year for health-related content regularly (black part of the bar) and occasionally (gray part of the bar) (2023).

How do websites and social media compare to other sources for health content? Perhaps not surprisingly, the most popular source of health information among our survey takers was medical professionals, with the majority (89%) reporting this experience. A close second were health websites at 77% and family and friends (73%), ahead of obtaining related content from the mainstream media (56%). Although considerably less common, a not insignificant number of people also reported government websites (41%), Wikipedia (30%), social media (29%), and video sites (19%) like YouTube and TikTok as health information sources. Overall, 82% of older adults reported obtaining health information from some type of digital source.

Returning to our recurring theme of highlighting inequality in who among older adults utilizes digital technologies, it is important to point out here that people with no more than a high school degree as well as people in more precarious financial situations were less likely to obtain health information from digital resources than others. There is no difference in age on this account, however; whether sixty or eighty, digital resources were equally accessed by older adults as sources of health information.

We shared earlier what proportion of older adults have heard of various online tools for learning. Now we look at how many of them actually use these resources. The most popular tool is online videos, with about four out of five older adults turning to it at least occasionally. More than half also turn to online references and Wikipedia occasionally, while about one in seven do so regularly. One-third learn from online discussion communities like Facebook Groups or Stack Overflow, and one-quarter from how-to websites. Experiences with formal online learning resources are the least common, with just 2% doing this regularly. The relative popularity of these online

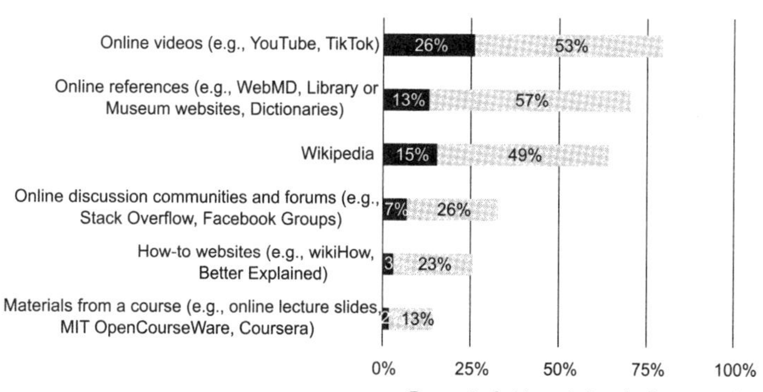

Popularity of various online learning tools among 2,000 older adults (darker shade signifies regular use of the resource, lighter shade occasional use) (2023).

resources mirrors those of other age groups, although all are more popular among younger people.[12]

One way older adults can contribute to others' learning is by sharing content themselves. We asked survey takers about their experiences with such activities. Half of them had posted a review online; the most popular such review was of a product other than a book, followed by a travel-related review and then book review. Plenty of older adults were clearly eager to share their opinions with the wider world. Many older adults were also engaged on online forums and social media through question asking and answering, with 41% having done this. A much smaller proportion (4% each) had edited something on Wikipedia or an online map, but the fact that 4% had done either or both shows that some older adults are very comfortable being actively engaged online in ways that are relatively rare among users of other ages as well.[13]

While clearly some older adults engage in online participa-

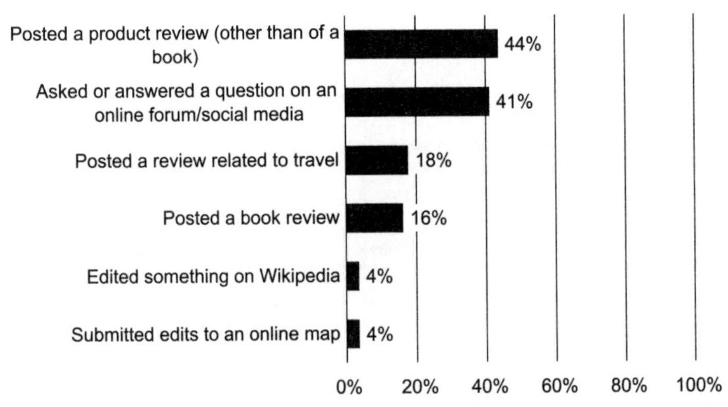

Proportion of 2,000 older adults having contributed various types of content online (2023).

tory activities, these numbers trail those of the more general adult population, leaving room for growth.[14] One study in the United States and one study in Spain both investigated the experiences of older adult bloggers and found that their contributions gave them continued purpose and also led to positive social connections, among other benefits.[15] Such results suggest worthwhile benefits to older adults contributing more online. Another type of online activity that could be beneficial to older adults is engagement in citizen science projects where volunteers without technical training can get involved through activities like labeling images to learn about the natural world (see Zooniverse for one example). Research has found that older adults tend to stick with such projects longer than their younger counterparts.[16] As noted throughout the book, lack of awareness of existing opportunities is the first barrier to using technologies. Once people know about what is possible, guidance in how to use the services can be valuable, especially if it comes in a kind and patient form.

Research about older adults and how they learn using new technologies offers a range of valuable clues that can guide those of us who care for older adults. These tips start with understanding the range of learning types and range of starting points among older adults; making the objectives of the learning clear and practical for older adults; and following through with plans for sustained support—which often fall by the wayside, despite the best of intentions. There are also a variety of different teaching and learning approaches that have been shown to work well for older adults that can be adapted for given types of learners, different settings, and varying goals.

A key starting point for those supporting older adults in their ongoing learning is to begin with a recognition of the diversity within this demographic and work from there. Older adults from older cohorts and with little office work experience are more likely to have lower general confidence in their ability to use new technologies as learners than those who are younger or who picked up technology skills in the workforce.[17] It is also likely that those older adults who have lower levels of educational attainment struggle more to use new technologies. But this does not mean that they do not have much to gain from informed usage. On the contrary, older adults, and those who struggle with the technologies, may well have the most to gain from thoughtful interventions and extra support.[18]

Instructors of older adults do well to make the goals of learning very clear at the outset. In addition to being clear about learning objectives of the course or the activity, it can be helpful for those who are trying to encourage older adults to learn to make the benefits of doing so obvious. Research shows that many older adults become more likely to learn the more they can see positive and *near-term* benefits that may come from their investment of time and energy. For instance, when older adults can imagine a practical use for the learning or the technology, their motivation to learn tends to increase. Examples tested in research studies include the benefits of using a tablet computer to take and share photographs with loved ones, accessing local services in a community that are only listed online, and exploring websites for health information.[19]

Research into how older adults learn using new technologies points to several approaches that teachers or supporters might use to help them. We highlight here three approaches that we consider especially promising: collaborative (or peer) learning, cooperative (or team-based) learning, and intergenerational learning.[20] While there are other teaching styles—also known as pedagogical approaches—we focus on these three that have strong research backing in each case.

Collaborative learning: Older adults benefit from working together with peers on learning, whether learning about new technologies or learning about other topics using those technologies. On the more formal end of the spectrum, the OLLI model that we explore in the opening vignette involving the Minow family tracks this aspect of the research. The use of peer coordinators who are proximate to the other learners both in age and in desire to learn can create the type of collaborative environment shown to work, according to systematic research. The OLLI model of using coordinators—who are typically, though not necessarily, older adults themselves—who support the learning of other older adults and learn alongside their peers is an example of this effective collaborative learning approach. On the informal end of the spectrum, this same approach of older adults learning collaboratively, intentionally learning from one another, can also take place effectively in less structured settings such as a social media environment and content creation.[21]

A related finding along these lines emerges from research that compares older adult learners to younger learners, such as those in their twenties. Older adult learners tend to be more collaborative and have learning goals that are more about gaining knowledge and insight, whereas younger learners are more likely to be competitive and sometimes driven by instrumental aims in their learning such as grades or career needs and aspirations.[22] This tendency among older adults to be more collaborative and less competitive is a useful clue for those who are supporting older adults in their learning to use new technologies.[23]

One study specifically highlighted that peer support can come from other community members like the learner and so-called super-users who already have more facility with the technology.[24] This study followed a handful of older adults for thirteen weeks to see whether weekly sessions with peers would facilitate the long-term adoption of smartphones. The researchers found that having a master-apprentice approach to learning was especially helpful because eventually everyone became a super-user, which meant that they could then serve as peer support for yet others in the community, making the learning process sustainable across a larger group.

We have continually emphasized the importance of seeing older adults as a heterogeneous mix of people rather than one monolith. To this point, it is important to note the findings from a study that examined long-term learning effects about e-health, comparing older adults with more and less technology experiences.[25] While those with more experiences benefited from collaborative learning, those with lower digital literacy did better long term when they had been offered individualized learning opportunities. To this end, ideally different learning contexts can be offered to cater to different learning needs.

Cooperative learning: A close cousin of collaborative learning, cooperative learning is a model in which older adults are assigned a task that is to be performed with others. In this model, those who are organizing the learning assign tasks that can be done in pairs or teams of older adults. The older adults work together, for instance, to read and discuss something, write an article, or solve a puzzle cooperatively. It is possible to combine approaches: a collaborative learning model may emphasize cooperative projects, while it is also possible for a cooperative project-based learning model to involve formal instructors who are older adults. This approach is also often combined with project-based learning, in which the learning takes place through a shared project that is assigned at the outset and carried out through the course or learning session. In our observation, this approach is better suited to the formal learning setting than to informal learning, given that the older adults can benefit from a well-crafted and guided cooperative learning prompt.

Intergenerational learning: Younger people can participate as teachers or peer learners alongside older adults. This approach works equally well in informal and formal learning settings. Intergenerational connections are consistently believed to be positive for the well-being and learning of older adults—and for younger people as well. Research repeatedly shows that older adults benefit from, and often like, learning from those they care about in younger generations, ideally in a comfortable setting such as the home and where the younger person is patient. As we discuss in several chapters, this support from younger people is especially helpful when setting up a device or first using an application or service. Face-to-face training and support are typically preferred by older adults where possible. Studies have specifically examined informal gaming-type and content creation environments as well as more formal learning settings.[26]

These approaches are not the only teaching and learning styles that can work for older adults. As we note throughout, there are some older adults who are terrific at learning on their own. While in the minority of older adults, these auto-didacts are in a great position. The internet offers such a huge range of options for learning on one's own, from videos on every imaginable topic to resources on how to teach oneself to use new technologies. AARP, for instance, has a very useful running list of books on how to use new technologies as an older adult, noting which books have won awards for effectiveness.[27]

Older adults are often turned off by the idea of learning about and through new technologies. Researchers have found that design considerations—making technologies more accessible for older adults as a core group of potential users—can make a big difference in adoption and usage. For instance, those who develop learning technologies would benefit from close coordination with educators who study and work with older adults as they develop new services. Getting feedback through focus groups and testing new tools with older adults may help as well. In any event, older adults should not be a forgotten constituency during the design process of new technologies.[28]

Key factors that jump out of the research for positive learning outcomes include easy and cost-effective access to the necessary technologies; internet access of a decent speed; and the relevant design and content of websites or applications.[29] Social factors also come into play, such as whether there are supportive younger people in the older adult's life who are encouraging and whether there are older adult peers who are engaging with similar technologies. If these technological and social factors are not in place for an older adult, their likelihood of starting or using new technologies for learning goes down. Said the other way around, by putting these elements in place, older adults are more likely to enjoy and benefit from positive learning outcomes using new technologies.

The good news is that these factors are achievable if we put our minds to the task and prioritize carrying it out equitably for older people in our lives and communities. In the concluding chapter that follows, we make recommendations for policy approaches that will ensure that older adults are well-served in their communities.

There is great value in older adults learning to use new technologies so that they can accomplish a range of goals. These goals might relate to personal interests they have, such as fly fishing or knitting, which they might want to look up. It might enable them to connect with loved ones who live far away or keep up with high school friends with whom they wish to remain close. It might serve basic human needs such as scheduling doctor's appointments, getting food delivery, or arranging a game of cards.

The most important outcome of learning to use new technologies might be all the other things that such aptitude can then make possible: in other words, the learning that can come from learning the technologies themselves. It is the classic "teach a person to fish rather than just giving them a fish" approach to empowering others.[30] Combined with strategies that encourage older adults to teach one another, we can set in motion a process that spreads learning and joy through older adult communities worldwide. Good policy in terms of digital inclusion starts from the premise that older adults are themselves participants in digital society—and can be powerful actors on behalf of themselves and others in their communities.[31]

This theme of learning about other topics through new technologies is one of the keys to successful aging. As we saw in the chapter on well-being, the use of new technologies among older adults can reduce the probability of a depressive state by as much as one-third. The work that the Minows and others like them have done through OLLI courses, online and in person, exemplify the types of activities that research shows may have this positive effect on older adults. By reducing isolation, activating curiosity, offering enjoyment from interactivity, and fostering connection between and among peers and generations, learning to use new technologies in cohort models can literally save and enrich the lives of older adults.

Research suggests this effect can be important for those younger than sixty as well. Those who are fifty and over, those beginning the "third chapter" of their life as some call it, have been included in some studies of this topic, showing once again that what is good for older adults is something that can benefit people of all ages.[32] Making sure that policies include the interests of older adults is crucial to achieving opportunities for all.

9

Lessons—For Older Adults, Their Families, Friends, and Society

We'll stress it again: older adults are not monolithic. There is extraordinary diversity within the population of older adults. By definition, as people live longer, the term "older adults" expands to include people at increasingly varied life stages. Not only are some older people still using modern technologies in very sophisticated ways at work, some of them even designed or invented the technologies. Likewise, there are some who are in their seventies, eighties, and nineties—as we have seen with Newton Minow—who are able and active in their use of new technologies daily.

Some older adults are still fit while others have faced health challenges for years. Varied circumstances also mean that people have diverse resources on which to draw both for financing their technology options and for getting assistance when they need it. These vast divergences must be kept in mind when thinking through how all older adults can thrive with technologies.

A most unfortunate stereotype about older adults is that they are universally clueless about technologies. This is not backed up by scientific evidence. Rather, there is considerable variation in their internet skills—just like there is in the digital literacy of other age groups—and not recognizing this short-changes the entire population. This stereotype can mean that older adults are not taken seriously as people who can learn and help themselves. It ignores the fact that older adults can help each other. Given that many older adults prefer to receive assistance from their peers, this is a major untapped resource for disseminating skills and assistance. And this myth makes it harder for younger people to learn from older adults—for example, in being more skeptical and critical of certain information that circulates online as we discuss in the chapters on safety and security and on misinformation. In sum, it is important to stop painting a picture of those in the later stages of life as not knowing what they are doing when it comes to technologies and assuming that they neither can nor want to learn. Many older adults are already quite savvy with technologies, and others are in a good position to be if given a chance through supportive assistance that comes with patience and respect. In some cases, they are even savvier than younger adults, such as in their ability to avoid falling for scams.

To address older adults' needs with technologies, here we outline how different actors can ensure equitable and effective access for older adults.

Advice for Older Adults

If you are an older adult who often needs help, remember that technologies are very forgiving, and you can try all sorts of approaches without breaking things. Trial and error is often a good first strategy to solving a problem. Many people who are savvy with devices got there through experimentation. If this does not work and your questions are not being addressed by those close in your social circles, think about the diverse resources that are available to you beyond your immediate surroundings. See whether local community centers or libraries offer workshops or drop-in clinic hours. Also remember that the internet itself can be a great source of assistance. Many YouTube and TikTok videos address a myriad of technology questions, from how to change the settings on your phone to how to fix issues with your printer.

The most recent advances in artificial intelligence are making it even easier to find answers to questions by offering the option of having humanlike exchanges with the system. While some of these require paid subscriptions, you can access others by setting up a free account. One such example is Google's Gemini, another is Microsoft's Copilot. At the prompt, the user can type in a question such as "how do i change the privacy settings of my facebook account" or "how do i turn off notification on my phone" and get step-by-step instructions from the service, often with visual aids. While writing such queries into search engines is also an option, these services give direct responses, and the user can also reply with additional questions. To be sure, these tools are very much in development, so users must be careful about trusting the responses they get. The systems often cite the sources of their answers, which the user can verify.

For actual human support instead of humanlike support, see whether any of your friends can provide assistance. Research has shown that many older adults are happy to help, but no one is asking them and so they end up being an untapped resource. Do not be shy about your questions; chances are others have also grappled with them.

If you are an older adult who is savvy and happy to help others, let your social circles know about this. In addition to doing a good deed by offering support, you can also learn new tips and tricks while giving assistance.

Advice for Those Who Care for Older Adults

Start early. We know that helping older adults begin using the technology in the first place can set them up for success. In addition to technological access—user-friendly devices, updated software, good and safe connectivity—older adults need access to support. If at all possible, be available to address the occasional question. Better yet, be proactive in offering suggestions on how things can be done and sharing new useful services you come across. The older adults in your life need to know that you care and that you are happy to help. The research shows that the best of intentions by support providers are not always followed up by helpful action on the part of those same well-intentioned persons. Following through is essential.

While often older adults tend to prefer in-person support, this is not always possible, and technology-mediated help can also be invaluable. Keep in mind old-fashioned technologies such as the telephone. A phone call to many older adults might be a great way to reconnect as well as to offer technological support while you are chatting away. Of course, online tools such as WhatsApp and FaceTime can work as well as the phone; by using these online tools to connect, the older adult will also gain confidence in using online tools in general.

Remember that the support person can be a peer (another older adult) or a younger person, whether one or two generations removed. The important aspect of the interaction is availability, patience, and respect. If it gets to be too much due to the frequency of requests, see whether you can split up the help with other caretakers. Another alternative is to look into what resources may exist in the older adult's community.

They may not think to check whether their local library offers technology sessions, but there is a good chance that it does. Consider accompanying them to such a session so that they can get through the initial barrier of using such a resource. Also see whether they have peers who could either help them or accompany them to such a session.

Advice for Public Libraries, Community Centers, and Senior Living Facilities

There is a good chance that staff at public libraries and community centers are well aware of the need for their services in the realm of technology support. In that case, consider using this book to make the case to your local government as to why your services are essential to supporting older adults in your community. Collaborating with local universities can also prove beneficial.[1]

In particular, the needed services concern broadly available and easily accessible support with technology questions that can range from hardware to software issues. The biggest barriers that older adults identify to getting connected concern security and privacy issues, so sessions focused on these would be especially welcomed. These are also topics that should likely permeate sessions on any other topical focus (see below for more on this).

There should be workshops targeted at different knowledge levels. There is, of course, need at the most basic level of signing up for an email account or using basic programs like word processing. But such organizations should not ignore the needs of those who are more knowledgeable yet still eager to learn and dig deeper into different topics.

Given the importance of online technologies for connecting to communities to avoid social isolation, sessions on informed social media use could offer considerable value. Signing up for a social media account, personalizing settings, and finding online communities of interest are just a few beneficial ways to help an older adult. Many users of Facebook do not realize that there are groups devoted to lots of hobbies and interests. Similarly, Reddit offers discussions of every topic under the sun, as we discuss in the chapter on learning. Even if focused on social connections, such a session should likely spend some time discussing the downsides of active online engagement like the possibility of being targeted for a scam. It is important to have such discussions in a way that does not scare away the user. Rather, offer guidance as to how to identify scams (recall our discussion of prebunking in the chapters on safety and security and on misinformation).

Another focus could be on how to find answers to technology (and other) questions online. Such a session could highlight question-and-answer sites like Quora, MrExcel, or Stack Overflow, but also Facebook Groups and Reddit as people often pose queries on these and receive helpful responses. It can also mention the many how-to videos that are available on services like YouTube and TikTok. Beyond pointing people to specific resources, such sessions could go a long way in helping attendees realize that whatever question they may have, someone else likely has already had it online and so an answer to it probably exists. And in case not, there are online communities where they can ask for assistance. This is a basic level of awareness that can really change someone's mindset about how to approach questions about technologies.

One approach to identify relevant topics for workshops is to ask community members what interests them. Nonetheless, this should not be the only approach as lacking awareness would prevent people from asking for assistance with things they may not even know they need or would enjoy. Thus, a mix of top-down and bottom-up choices in topics is likely best.

Advice for Technology Developers

Older adults are interested in technologies, but sometimes have unique needs such as accessibility features due to decreasing mobility and eyesight. Technology developers need to keep these in mind as they create and update devices and services. There is a clear incentive here: making services inclusive to users from a segment of the population with more time and money than most. It is fine to cater to younger people to set them up as lifelong clients, but it is foolish to ignore the needs of users who are in the best position to spend money on technology resources. Include older adults in your planning and development, and also in your advertisements as potential users, so it is clear to them that they are part of your constituents.

Advice for Companies

Companies not in the technology sector also have lessons to learn when it comes to older adults' digital media uses. Whatever strategies such corporations use for reaching their clients, using technologies should be inclusive of older adults. When developing an online marketing campaign, keep in mind that adults of all ages may see your materials. Be sure that the content is inclusive of different ages. Many older adults are on Facebook, fewer are on Instagram and especially TikTok, but they are not foreign to such platforms either. Exclude them at your peril.

Advice for Journalists

Press coverage of older adults and technology use too often portrays this population as lost and confused. Instead of painting a broad brush of older adults as uniformly unskilled, it is also important to highlight the experiences of those who are savvy and benefiting from their technology uses. Additionally, the common fearmongering about social media is likely to frighten potential users rather than draw them in. Given scholarly evidence that social media use can be beneficial, it behooves journalists to cover this domain with more care.

Advice for Policymakers and All Who Vote

Older adults vary widely in how they have embraced new technologies. Many, for reasons we discuss in the book, are afraid or disinterested in getting started with new technologies. Others cannot afford access, or low levels of literacy make it nearly impossible to get going with new technologies. The range in terms of sophistication and participation in this new world is growing greater, not less great, with the advent of new technologies. This development itself is a policy problem—it is yet another way in which the divide between the "haves" and "have-nots" is growing at a global scale.

Communities that seek to support and engage their older adults would do well to consider policies that are older-adult-friendly when it comes to technology. Focus on the "last mile" for technologies to ensure that older adults can gain access to decent broadband at reasonable cost, free services at venues such as libraries and community centers, and training opportunities with people equipped to help them get started and who can offer support along the way. These services cannot be accessible only to the wealthy, those in urban or suburban areas, and those living in the Global North if we want to ensure a more equitable, inclusive future in the digital era.

Provide specific, ongoing opportunities for learning for older adults using technologies designed to include older adults, noting the range of ability and disability is greater than in other populations. Too many districts are cutting funding to public libraries thinking they are part of a bygone era. Nothing could be further from the truth. Libraries have huge roles to play in providing support to people of all ages with their technology needs. But they can do this only if they have the personnel needed to staff technology workshops and clinic hours.

In the realm of safety and security—one of the major barriers for some older adults to expanding their online actions—policymakers have an important role to play in developing more extensive and older-adult-friendly provisions for consumer protection. Ensure that older adults have the ability to gain recourse when they are harmed online or to switch services when they feel the need to do so. Policies should encourage the availability of data portability and ensure interoperability between systems to make things easy for older adults when they want to end the use of a service.

Privacy is another area where older adults express concerns (as do adults of all ages). They need training support to gain an awareness of their options and to set up the devices and services they use in ways that reflect their preferences. They also need easy places to turn to when something goes wrong, and law enforcement needs the tools and the resources to help them. At an individual level, law enforcement needs to be able to address the growing number of online scams that target older adults. At the systemic level, policymakers need to be able to hold technology companies and platforms accountable for providing a reasonably safe and secure system for older adults to use without fear.

Why Focusing on Older Adults and Their Technology Use Matters

Older adults matter in every society, in every family. We should honor, support, and cherish older adults not just for what they have already given, but for what they can still contribute. Too often, we forget that older adults are key actors in the world.

Reviewing the research literature about older adults and digital media uses in 2018, it was shocking to see just how little scholarship had gone beyond whether older adults were online at all.[2] Compared to what we knew then, we know considerably more now, which is why writing this book was possible. There is valuable research available now on both the opportunities and challenges for older adults using digital technologies, which served as the basis for this book. Nonetheless, a tremendous amount remains unknown both because it is an ever-changing landscape and because the focus on older adults in the realm of internet research has been slow to develop. This is sadly in line with older adults' perception of feeling invisible in everyday life.

Much more will need to be done in this domain to understand from a more holistic perspective what the health implications are of various online activities or which particular online behaviors contribute to people feeling socially connected versus isolated. As technologies like generative AI develop, we also need to know more about the extent to which people understand manipulated images and videos as these may have considerable implications for the democratic process and other key areas of life. These questions exist for the population as a whole, but too often older adults are excluded from such studies. They are just as important a population as any other to study.

The use of new technologies can be a key element of successful aging, and some older adults already embrace these innovations in many aspects of their lives. Everyone needs to be able to access the health, financial, and civic information that is now predominantly online. In instrumental terms, greater technological access and savvy can be a major social driver ensuring that older adults contribute more to civic and economic life for a longer period. At a personal level, we should see digital technologies as a way to ensure greater well-being for those we love and ourselves.

10

Top Ten Takeaways

Here are ten key takeaways from the research on older adults and their use of new technologies:

1. The use of new technologies by older adults can be a key component of successful aging. There are huge benefits for older adults to using new technologies to connect to other people and to learn new things. They face real risks to their safety and privacy too. We all need to focus on both sides of this coin.
2. Contrary to popular belief, many older adults are more capable than we give them credit for when it comes to new technologies. While they need support, like everyone, we should not view older adults as a monolith when it comes to using new technologies.
3. Older adults are surprisingly agile in switching between technology services. (They go where their family and friends are.) We need to be sure that it is easy for them to have a sense of control and to be able to bring their data with them.
4. Younger family members and friends offering support to older adults can result in gains for everyone involved—for the members of multiple generations. Younger people need to be sure to follow through when offering support!

5. Older adults are happy to help others with technology questions. (It's just that too often, people don't think to ask older adults for help!) Older adults often prefer to get help from their peers. (Peers tend to be more patient with them than younger people.)
6. Meaningful connections online among older adults, and those with other generations, can be important for counteracting loneliness. (Superficial scrolling may have the opposite effect.) The possibilities for lifelong learning and connection to new friends are endless for older adults—and can be a lifeline when it is harder to get around physically or unsafe to do so from a health perspective.
7. Many older adults are less likely, not more likely, to fall for scams than younger adults. What sets older adults apart from younger people is that *they are targeted more often*, and for some, cognitive decline can get in the way of making sound decisions. These findings are true both online and offline. Scams sometimes involve both.
8. Investments in support for older adults in their use of new technologies can pay huge social benefits. Companies that develop technologies and those who design them often underinvest in the needs of older adults—to everyone's detriment.
9. Too often older adults are left out of the conversation about new technologies—both in terms of the real risks they face and the benefits they can gain, such as from online courses or health information.
10. It is a mistake to leave technology policy to technology companies, as we have done in the United States over the past twenty-five years. The needs of older adults—and others—require focused policy and investments in technology as in all areas of law and policy. Leaving it "to the market" once again will lead to terrible harms and missed opportunities for older adults and for society as a whole.

ACKNOWLEDGMENTS

Our thinking on the topic of the book has benefited from conversations and collaborations with several people over the years. We thank Catherine Bracy, Sandra Cortesi, Urs Gasser, Amanda Hunsaker, Minh Hao Nguyen, Eric Sears, and Nicol Turner Lee for many helpful exchanges. Becca Smith and Annalise Baines made important contributions to the survey study. Amanda Hunsaker, Minh Hao Nguyen, Teodora Djukaric, Jaelle Fuchs, and Gökçe Karaoglu contributed to the interview study. Emilia Volkart, Kate Blinova, and Rabiya Abdullah helped with identifying and summarizing relevant literature. Balázs Hargittai, István Hargittai, Magdolna Hargittai, Martha Minow, Mary Minow, Will Marler, and Annalise Baines kindly read all or parts of the manuscript's earlier versions and offered insightful comments. We appreciate all of these contributions as well as the helpful input we received from four anonymous reviewers.

Eszter is deeply grateful to the daily inspiration that her writing accountability group offers and sends big thanks to Maria Apostolaki, Agnes Bäker, and Erin Kelly. She did some of the work on this book while holding the William Allan Neilson Professorship at her alma mater Smith College and deeply appreciates the opportunity for research and reflection that visit enabled. She also feels extremely fortunate to get the kind of support for research that the University of Zurich offers.

Joseph Calamia has been an incredibly helpful editor, and the book is all the better for it. We thank Joe for his many great suggestions and all the energy he poured into the project. We also appreciate the valuable input we received from copy-editor Lisa A. Wehrle.

A grant from the John D. and Catherine T. MacArthur Foundation enabled the collection of original survey data on which we rely throughout the text. A grant from the Templeton Religion Trust supported a data set we draw on in chapter 2.

Whether through formal interviews or informal conversations, we have learned from hundreds of older adults about their experiences with, hopes for, and concerns about new technologies. We thank them for taking the time to share their insights with us and inspiring the material in this book.

METHODOLOGICAL APPENDIX

Survey of Older Adults

This section gives details about the survey that forms the basis of original findings we share in the various chapters. Data collection took place between September 21 and October 10, 2023. We surveyed 4,000 adults sixty and older split between two surveys of 2,000 each. The reason for this design was that we wanted to ask more questions than is reasonable for one survey. Long surveys result in poor response quality as they are more cognitively taxing and respondents get fatigued. To avoid this problem, we administered two separate surveys with some questions asked of both groups, but many questions asked from only one or the other sample.

We contracted with the survey firm YouGov to reach respondents from their US survey panel. YouGov maintains a panel with millions of members representing diverse backgrounds. Participants are compensated for their time. The platform is accessible, complying with Web Content Accessibility Guidelines (WCAG 2.1) including keyboard navigation and screen reader support plus voice controls on mobile apps, making it possible for people with varying abilities to participate in the surveys.

It is important to keep in mind that older adults who take a survey online are likely to be different from the average older adult due to more digital opportunities (more devices, better connectivity, better skills) to engage online in that way. To address this discrepancy, we apply survey weights to the data. These were provided by YouGov based on age, gender, race/ethnicity, and education, all of which are general predictors of internet uses and contexts and so would account for some of the possible sample divergences. The weights of 90% of the cases are between 0.60–1.67; overall, weights range from 3.0–7.0.

A limitation of surveys more generally is that people may be responding mechanically without really considering the questions or their answers. Since people get financial compensation for taking surveys, they may prefer to skip through questions quickly to get the payment and move on rather than taking them seriously. This poses potential quality concerns. To address these, researchers can use various methods to check whether participants are paying attention. There is considerable scholarship on what methods work best, but also on how such checks can ultimately bias against certain participants.[1] Keeping all this in mind, we implemented one attention-check question. Everybody included in the analyses answered this question correctly. Additionally, YouGov uses "natural language processing to identify and exclude poor quality responses."[2]

Table A.1 shows the sociodemographic makeup of the sample of 4,000 people surveyed for this book, both unweighted and weighted as per the explanation above. Table A.2 shows the distribution of well-being measures for the weighted sample. Table A.3 displays information about the digital experiences of the weighted sample.

Table A.1. Sample characteristics of the 4,000 older adults in the 2023 survey, unweighted and weighted (percentages)

	Unweighted	Weighted
Age		
60–64	25.2	23.5
65–69	26.5	27.1
70–74	20.9	19.9
75–79	15.9	15.8
80–94	11.6	13.7
Female	54.2	54.6
Race and ethnicity		
White	76.1	74.8
Black	9.9	10.0
Asian	2.3	2.2
Native	0.7	0.6
Hispanic	9.8	11.2
Education		
High school or less	39.8	42.8
Some college	26	25.1
College or more	34.3	32.0
Financial hardships		
0	66.8	66.2
1	19.6	19.9
2	10.1	10.1
3	3.6	3.8
Residence		
Urban	45.1	44.7
Suburban	34.3	34.2
Rural	20.6	21.1
Disabled	26.1	26.5
Employed	21.4	20.4

Table A.2. The distribution of well-being measures of the weighted older adults survey sample (N = 4,000)

	Mean
Life satisfaction (0–100)	72.9
Happiness (1–3)	2.2
Social connectedness (1–6)	4.7
Health (1–5)	3.2
Eudaimonic well-being (1–5)	3.3
Anxiety (1–4)	1.7
Depression (1–4)	1.8

Table A.3. The digital experiences of older adult survey participants based on weighted analyses (percentages, N = 4,000)

	Mean
Has access to a mobile phone	96.4
Has access to a tablet	88.6
Has access to a computer	97.1
Has access to all three	85.5
Goes online regularly	89.3
General internet skills (1–5)	2.8
Social media skills (1–5)	3.0

WELL-BEING MEASURES

The survey asked about seven types of well-being: life satisfaction, happiness, social connectedness, health, eudaimonic well-being, anxiety, and depression. This section explains how we measured each of them, all derived from and informed by expert literature on the topic.

To assess life satisfaction, we asked: "Taking all things together, how satisfied are you with your life as a whole these days? Indicate on this slider from 0 to 100 where zero is very dissatisfied and 100 is very satisfied,"[3] resulting in a continuous measure from 0 to 100.

To measure happiness, we asked: "In general, how happy would you say you are—very happy, fairly happy, or not too happy?" with respondents picking one of those three options.[4] This resulted in a three-point scale.

For social connectedness, we relied on a measure by Lee and Robbins[5] that presents respondents with a list of statements and asks them to what extent they agree with them on a six-point scale. The instructions say: "Below are some statements about how people might feel. When thinking about the past two weeks, please indicate how much you disagree or agree with the statements. There are no right or wrong answers. Try to not spend too much time on any one statement but give the answer which seems to describe your feelings best." The statements are:

- I felt disconnected from the world around me.
- Even around people I know, I didn't feel that I really belonged.
- I felt so distant from people.
- I had no sense of togetherness with my peers.
- I caught myself losing all sense of connectedness with society.
- I didn't feel I participated with anyone or any group.

We average the agreement score of each statement to end up with a mean that can range from 1-6.

To gauge people's health status, we asked: "In general, how is your health?" with five possible response options: excellent, very good, good, fair, poor,[6] resulting in a 1–5 scale.

To get a sense of people's eudaimonic well-being, the survey offered the following instructions: "In this question, please reflect on the past month. How did you perceive your activities and experiences in the past month?"[7] This was asked concerning the following sentiments:

- . . . meaningful to me
- . . . valuable to me
- . . . precious to me
- . . . full of significance to me
- . . . fulfilling to me

The answer options were: does not apply at all, applies a little, applies somewhat, applies a lot, applies fully. We averaged the scores resulting in a 1–5 scale.

For assessing anxiety, we asked respondents to share how they feel using the following prompt: "Below are some statements about how people sometimes feel. Please indicate how often you felt that way during the past week. The best answer is usually the one that comes to your mind first, so do not spend too much time on any one statement."[8] These were the five feelings listed:

- I had fear of the worst happening.
- I was nervous.
- I felt my hands trembling.
- I had a fear of dying.
- I felt faint.

Respondents were asked to pick one of the following for each: never, hardly ever, some of the time, most of the time. We averaged these resulting in a 1–4 scale.

We gave similar instructions for assessing depression and again asked survey participants to indicate the frequency with which they experienced a list of feelings.[9] The prompt read: "Below is a list of the ways you might have felt or behaved. Please mark how often you have felt this way during the past week." The experiences listed were:

- I felt depressed.
- I felt that everything I did was an effort.
- My sleep was restless.
- I was happy.
- I felt lonely.
- I enjoyed life.
- I felt sad.
- I could not get "going."

The frequency options were:

- Rarely or none of the time (less than 1 day)
- Some or a little of the time (1–2 days)
- Occasionally or a moderate amount of time (3–4 days)
- Most or all of the time (5–7 days)

After reverse-coding "I felt happy" and "I enjoyed life," we averaged the responses to get a 1–4 scale for the depression score.

Survey of Adults' Social Media Uses

The chapter on adoption contains two graphs where we compare the social media experiences of older adults with younger adults. This section describes details about that data set, which is entirely independent of the older adults survey described earlier. We conducted this study as part of a different project on "Learning about Science and Religion Online: Who, What, Where, How?" supported by a grant to Eszter from a joint project of Rice University and the University of California, San Diego, funded by the Templeton Religion Trust.

We administered this survey in June 2023, contracting with the survey research organizations YouGov and NORC AmeriSpeak. The aim was a nationally representative sample. Both survey companies provided weights based on age, gender, education, and race/ethnicity, which we employed in the analyses. The sample includes responses from 2,505 participants, all of whom passed an attention-check question in addition to other indicators of valid responses. Table A.4 shows the characteristics of the June 2023 survey sample with data about social media experiences.

Table A.4. Sample characteristics of the June 2023 survey about social media experiences

	Unweighted		Weighted		
	Percent	Mean	Percent	Mean	N
Age		48.7		48.2	2,505
Female	51.7		51.2		2,505
Race and ethnicity					2,486
White	66.6		62.7		
Black	13.0		13.5		
Asian	3.7		5.3		
Native	1.6		1.6		
Hispanic	14.0		15.9		
Education					2,505
High school or less	29.8		38.6		
Some college	34.3		26.7		
College or more	24.4		23.1		
Household income		76,111.1		75,474.0	2,493
Residence					2,486
Urban	50.3		48.9		
Suburban	32.7		34.6		
Rural	17.0		16.2		
Disabled	16.4		16.9		2,496

Interview Study of Older Adults

We conducted over 100 semi-structured in-depth interviews with older adults in Bosnia, Hungary, the Netherlands, Serbia, Switzerland, Turkey, and the United States between December 2018 and March 2020 (before the COVID-19 pandemic). The ages of respondents ranged from just shy of sixty years old to ninety-one. We conducted the majority of interviews in person with the exception of sixteen of the thirty-six US cases, which took place via phone or video call. In-person interviews took place at participants' homes or at an agreed-upon location (e.g., café, library). Interviews were mostly in the local language, although in some cases when respondents preferred, we spoke in English even if that was not the local language. The length of the interviews ranged from 8 to 135 minutes depending on a participant's level of digital experiences; the average interview lasted about an hour. Participants received financial compensation for their participation: 20 BAM, 3,000 HUF, 20 EUR, 1,000 RSD, 30 CHF, 40 TRY, or 20 USD, respectively, depending on the country of the interview, taking into account purchasing parity across locations.

NOTES

CHAPTER 1

1. United Nations, "World Population Prospects: The 2017 Revision, Key Findings and Advance Tables" (Department of Economic and Social Affairs, Population Division, 2017); World Health Organization, "Ageing," World Health Organization, accessed May 31, 2024, https://www.who.int/health-topics/ageing#tab=tab_1.

2. John W. Rowe and Robert L. Kahn, "Successful Aging," *Gerontologist* 37, no. 4 (August 1, 1997): 433–40, https://doi.org/10.1093/geront/37.4.433; John W. Rowe and Robert L. Kahn, "Successful Aging 2.0: Conceptual Expansions for the 21st Century," *Journals of Gerontology: Series B* 70, no. 4 (July 1, 2015): 593–96, https://doi.org/10.1093/geronb/gbv025; Alexandra M. Freund and Michaela Riediger, "Successful Aging," in *Handbook of Psychology*, vol. 6, *Developmental Psychology*, ed. Richard M. Lerner, M. Ann Easterbrooks, and Jayanthi Mistry, 2nd ed. (Hoboken, NJ: John Wiley & Sons, 2003), https://www.wiley.com/en-us/Handbook+of+Psychology%2C+Volume+6%2C+Developmental+Psychology%2C+2nd+Edition-p-9780470768860; George E. Vaillant, *Aging Well: Surprising Guideposts to a Happier Life from the Landmark Harvard Study of Adult Development* (New York: Little, Brown, 2002); G. E. Vaillant and K. Mukamal, "Successful Aging," *American Journal of Psychiatry* 158, no. 6 (2001): 839–47, https://doi.org/10.1176/appi.ajp.158.6.839.

3. danah boyd, *It's Complicated: The Social Lives of Networked Teens* (New Haven, CT: Yale University Press, 2014); Nicole B. Ellison, Charles Steinfield, and Cliff Lampe, "The Benefits of Facebook 'Friends': Social Capital and College Students' Use of Online Social Network Sites," *Journal of Computer-Mediated Communication* 12, no. 4 (2007): 1143–68, https://doi.org/10.1111/j.1083-6101.2007.00367.x; Mizuko Ito et al., *Hanging Out, Messing Around, and Geeking Out: Kids Living and Learning with New Media* (Cambridge, MA: MIT Press, 2009); H. Jenkins, M. Ito, and d. boyd, *Participatory Culture in a Networked Era* (Cambridge, UK:

Polity Press, 2016); S. Livingstone, *Children and the Internet* (Cambridge, UK: Polity Press, 2009); John Palfrey and Urs Gasser, *Born Digital: Understanding the First Generation of Digital Natives* (New York: Basic Books, 2008); S. Craig Watkins, *The Young and the Digital: What the Migration to Social Network Sites, Games, and Anytime, Anywhere Media Means for Our Future* (Boston: Beacon Press, 2009).

4. Teresa Correa, "The Participation Divide Among 'Online Experts': Experience, Skills and Psychological Factors as Predictors of College Students' Web Content Creation," *Journal of Computer-Mediated Communication* 16, no. 1 (2010): 71–92, https://doi.org/10.1111/j.1083-6101.2010.01532.x; Eszter Hargittai, "Digital Na(t)Ives? Variation in Internet Skills and Uses among Members of the 'Net Generation'*," *Sociological Inquiry* 80, no. 1 (2010): 92–113, https://doi.org/10.1111/j.1475-682X.2009.00317.x; Eszter Hargittai and Amanda Hinnant, "Digital Inequality: Differences in Young Adults' Use of the Internet," *Communication Research* 35, no. 5 (2008): 602–21, https://doi.org/10.1177/0093650208321782; Eszter Hargittai and G. Walejko, "The Participation Divide: Content Creation and Sharing in the Digital Age," *Information, Communication and Society* 11, no. 2 (2008): 239–56, https://doi.org/10.1080/13691180801946150; Sonia Livingstone and Ellen Helsper, "Gradations in Digital Inclusion: Children, Young People and the Digital Divide," *New Media & Society* 9, no. 4 (2007): 671–96, https://doi.org/10.1177/1461444807080335; Siva Vaidhyanathan, "Generational Myth," *Chronicle of Higher Education* 55, no. 4 (2008): B7–9.

5. Heinz Bonfadelli, "The Internet and Knowledge Gaps: A Theoretical and Empirical Investigation," *European Journal of Communication* 17, no. 1 (2002): 65–84, https://doi.org/10.1177/0267323102017001607; Kerry Dobransky and Eszter Hargittai, "The Disability Divide in Internet Access and Use," *Information, Communication and Society* 9, no. 3 (2006): 313–34, https://doi.org/10.1080/13691180600751298; R. Eynon, "Mapping the Digital Divide in Britain: Implications for Learning and Education," *Learning, Media, and Technology* 34, no. 4 (2009): 277–90, https://doi.org/10.1080/17439880903345874; Donna L. Hoffman, Thomas P. Novak, and Ann Schlosser, "The Evolution of the Digital Divide: How Gaps in Internet Access May Impact Electronic Commerce," *Journal of Computer-Mediated Communication* 5, no. 3 (2000), https://doi.org/10.1111/j.1083-6101.2000.tb00341.x; Hiroshi Ono and Madeline Zavodny, "Gender and the Internet," *Social Science Quarterly* 84, no. 1 (2003): 111–21; Neil Selwyn, Stephen Gorard, and John Furlong, "Whose Internet Is It Anyway? Exploring Adults' (Non)Use of the Internet in Everyday Life," *European Journal of Communication* 20, no. 1 (2005): 5–26, https://doi.org/10.1177/0267323105049631.

6. Grant Blank and Bianca C. Reisdorf, "The Participatory Web," *Information, Communication & Society* 15, no. 4 (2012): 537–54, https://doi.org/10.1080/1369118x.2012.665935; Paul DiMaggio et al., "Digital Inequality: From Unequal Access to Differentiated Use," in *Social Inequality*, ed. Kathryn Neckerman (New

York: Russell Sage Foundation, 2004), 355–400; William H. Dutton and Grant Blank, "The Emergence of Next Generation Internet Users," *International Economics and Economic Policy* 11, no. 1–2 (2014): 29–47, https://doi.org/10.1007/s10368-013-0245-8; Eszter Hargittai, "Second-Level Digital Divide: Differences in People's Online Skills," *First Monday* 7, no. 4 (April 1, 2002), http://firstmonday.org/ojs/index.php/fm/article/view/942; Karen Mossberger, Caroline J. Tolbert, and Mary Stansbury, *Virtual Inequality: Beyond the Digital Divide* (Washington, DC: Georgetown University Press, 2003); Pippa Norris, *Digital Divide: Civic Engagement, Information Poverty and the Internet in Democratic Societies* (New York: Cambridge University Press, 2001); Bianca C. Reisdorf and Darja Groselj, "Internet (Non-)Use Types and Motivational Access: Implications for Digital Inequalities Research," *New Media & Society* 19, no. 8 (August 1, 2017): 1157–76, https://doi.org/10.1177/1461444815621539; Jen Schradie, "The Digital Production Gap: The Digital Divide and Web 2.0 Collide," *Poetics* 39, no. 2 (2011): 145–68, https://doi.org/10.1016/j.poetic.2011.02.003; Neil Selwyn, "Defining the 'Digital Divide': Developing a Theoretical Understanding of Inequalities in the Information Age" (Occasional Paper 49, School of Social Sciences, Cardiff University, 2002), http://ictlogy.net/bibliography/reports/projects.php?idp=348; Neil Selwyn and Keri Facer, *Beyond the Digital Divide: Rethinking Digital Inclusion for the 21st Century* (Bristol: FutureLab, 2007); M. Warschauer, "Reconceptualizing the Digital Divide," *First Monday* 7, no. 7 (2002), http://dx.doi.org/10.5210/fm.v7i7.967.

7. Moritz Büchi and Eszter Hargittai, "A Need for Considering Digital Inequality When Studying Social Media Use and Well-Being," *Social Media + Society* 8, no. 1 (January 1, 2022): 20563051211069125, https://doi.org/10.1177/20563051211069125; Eszter Hargittai, "The Digital Reproduction of Inequality," in *Social Stratification*, ed. David Grusky (Boulder, CO: Westview Press, 2008), 936–44; Eszter Hargittai, *Connected in Isolation: Digital Privilege in Unsettled Times* (Cambridge, MA: MIT Press, 2022); Alexander J. A. M. van Deursen et al., "The Compoundness and Sequentiality of Digital Inequality," *International Journal of Communication* 11 (2017): 22.

8. Mary Christine Cremin, "Feeling Old versus Being Old: Views of Troubled Aging," *Social Science & Medicine* 34, no. 12 (June 1, 1992): 1305–15, https://doi.org/10.1016/0277-9536(92)90139-H; Newgate Research, *State of the (Older) Nation 2018: A Nationally Representative Survey Prepared for the COTA Federation (Councils on the Ageing)* (Sydney: Newgate Research, 2018), https://cnpea.ca/images/cota-state-of-the-older-nation-report-2018-final-online.pdf; Laura Kadowaki, Barbara McMillan, and Kahir Lalji, *Consultations on the Social and Economic Impacts of Ageism in Canada: "What We Heard" Report* (Federal, Provincial and Territorial (FPT) Forum of Ministers Responsible for Seniors, 2023), https://www.canada.ca/content/dam/esdc-edsc/documents/corporate/seniors/forum/reports/consultation-ageism-what-we-heard/UWBC-FinalDraftReport

-WWH-EN-20230411-withISBN.pdf; Kelly McDougall and Helen Barrie, *Retired Not Expired* (Adelaide: University of Adelaide, 2020), https://www.sahealth.sa.gov.au/wps/wcm/connect/d7ef0d6e-9157-499d-a373-e3c7d249f685/Retired+Not+Expired+Report.pdf?MOD=AJPERES&CACHEID=ROOTWORKSPACE-d7ef0d6e-9157-499d-a373-e3c7d249f685-nwKW63c; Sandra L. McGuire, Diane A. Klein, and Shu-Li Chen, "Ageism Revisited: A Study Measuring Ageism in East Tennessee, USA," *Nursing & Health Sciences* 10, no. 1 (2008): 11–16, https://doi.org/10.1111/j.1442-2018.2007.00336.x; Victor Minichiello, Jan Browne, and Hal Kendig, "Perceptions and Consequences of Ageism: Views of Older People," *Ageing & Society* 20, no. 3 (May 2000): 253–78, https://doi.org/10.1017/S0144686X99007710; Office for Seniors, *Attitudes towards Ageing: Research Commissioned by the Office for Seniors* (Wellington, NZ: Office for Seniors, 2016), https://officeforseniors.govt.nz/assets/documents/our-work/Ageing-research/Attitudes-towards-ageing-summary-report-2016.pdf; Erdman B. Palmore, "Research Note: Ageism in Canada and the United States," *Journal of Cross-Cultural Gerontology* 19, no. 1 (March 2004): 41–46, https://doi.org/10.1023/B:JCCG.0000015098.62691.ab; Sue Westwood, "'It's the Not Being Seen That Is Most Tiresome': Older Women, Invisibility and Social (in)Justice," *Journal of Women & Aging* 35, no. 6 (November 2, 2023): 557–72, https://doi.org/10.1080/08952841.2023.2197658.

9. A. Patricia Aguilera-Hermida, "Fighting Ageism through Intergenerational Activities, a Transformative Experience," *Journal of Transformative Learning* 7, no. 2 (November 30, 2020): 6–18; Michael D. Barnett and Cassidy M. Adams, "Ageism and Aging Anxiety among Young Adults: Relationships with Contact, Knowledge, Fear of Death, and Optimism," *Educational Gerontology* 44, no. 11 (November 2, 2018): 693–700, https://doi.org/10.1080/03601277.2018.1537163; David Burnes et al., "Interventions to Reduce Ageism Against Older Adults: A Systematic Review and Meta-Analysis," *American Journal of Public Health* 109, no. 8 (August 2019): 1–9, https://doi.org/10.2105/AJPH.2019.305123; Lisbeth Drury, Dominic Abrams, and Hannah J. Swift, *Making Intergenerational Connections—An Evidence Review: What Are They, Why Do They Matter and How to Make More of Them* (London: Age UK, 2017), https://www.ageuk.org.uk/globalassets/age-uk/documents/reports-and-publications/reports-and-briefings/active-communities/rb_2017_making_intergenerational_connections.pdf; Divya Raina and Geeta Balodi, "Ageism and Stereotyping of the Older Adults," *Scholars Journal of Applied Medical Sciences* 2 (January 1, 2014): 733–39.

10. Neil Charness, "A Framework for Choosing Technology Interventions to Promote Successful Longevity: Prevent, Rehabilitate, Augment, Substitute (PRAS)," *Gerontology* 66, no. 2 (2020): 169–75, https://doi.org/10.1159/000502141; Sara J. Czaja et al., "Older Adults and Technology Adoption," *Proceedings of the Human Factors and Ergonomics Society Annual Meeting* 52, no. 2 (September 1, 2008): 139–43, https://doi.org/10.1177/154193120805200201; Sara J. Czaja et al.,

Designing for Older Adults: Principles and Creative Human Factors Approaches, 3rd ed. (Boca Raton, FL: CRC Press, 2019), https://doi.org/10.1201/b22189.

11. United Nations, "World Population Prospects 2019" (United Nations Population Division, 2019), https://population.un.org/wpp/Graphs/Probabilistic/POP/60plus/900; World Health Organization, "The Determinants of Health," World Health Organization, Health Impact Assessment, 2018, http://www.who.int/hia/evidence/doh/en/.

12. Monica Anderson and Andrew Perrin, *Tech Adoption Climbs Among Older Adults* (Washington, DC: Pew Research Center, 2017), http://www.pewinternet.org/2017/05/17/tech-adoption-climbs-among-older-adults/.

13. Michelle Faverio, "Share of Those 65 and Older Who Are Tech Users Has Grown in the Past Decade," *Pew Research Center* (blog), January 13, 2022, https://www.pewresearch.org/short-reads/2022/01/13/share-of-those-65-and-older-who-are-tech-users-has-grown-in-the-past-decade/.

14. Eszter Hargittai, ed., *Handbook of Digital Inequality* (Cheltenham, UK: Edward Elgar, 2021); Laura Robinson et al., "Digital Inequalities and Why They Matter," *Information, Communication & Society* 18, no. 5 (2015): 569–82, https://doi.org/10.1080/1369118X.2015.1012532.

15. Anderson and Perrin, *Tech Adoption*.

16. Rowe and Kahn, "Successful Aging."

17. Rowe and Kahn, 433.

18. Urtamo Annele, K. Jyväkorpi Satu, and E. Strandberg Timo, "Definitions of Successful Ageing: A Brief Review of a Multidimensional Concept," *Acta Bio Medica: Atenei Parmensis* 90, no. 2 (2019): 359–63, https://doi.org/10.23750/abm.v90i2.8376.

19. Galit Nimrod, "Aging Well in the Digital Age: Technology in Processes of Selective Optimization with Compensation," *Journals of Gerontology: Series B* 75, no. 9 (October 16, 2020): 2008–17, https://doi.org/10.1093/geronb/gbz111.

20. Hargittai, "Second-Level Digital Divide"; Alexander J. A. M. van Deursen and J. A. G. M. van Dijk, "Internet Skills and the Digital Divide," *New Media & Society* 13, no. 6 (2011): 893–911, https://doi.org/10.1177/1461444810386774.

21. Kelly Marnfeldt et al., "'Connect It Down to the Person': Perspectives on Technology Adoption from Older Angelenos," *Journal of Elder Policy* 2, no. 3 (2023): 93–126, https://doi.org/doi: 10.18278/jep.2.3.4; Constance Elise Porter and Naveen Donthu, "Using the Technology Acceptance Model to Explain How Attitudes Determine Internet Usage: The Role of Perceived Access Barriers and Demographics," *Journal of Business Research* 59, no. 9 (2006): 999–1007, https://doi.org/10.1016/j.jbusres.2006.06.003; Reisdorf and Groselj, "Internet (Non-)Use Types"; Tanja Schroeder et al., "Older Adults and New Technology: Mapping Review of the Factors Associated with Older Adults' Intention to Adopt Digital Technologies," *JMIR Aging* 6, no. 1 (May 16, 2023): e44564, https://doi.org/10.2196/44564; Mark Tyler, Linda De George-Walker, and Veronika Simic, "Motivation

Matters: Older Adults and Information Communication Technologies," *Studies in the Education of Adults* 52, no. 2 (July 2, 2020): 175–94, https://doi.org/10.1080/02660830.2020.1731058.

22. State of Illinois, "Articles of Incorporation: John D. and Catherine T. MacArthur Foundation" (State of Illinois, Office of the Secretary of State, 1970).

23. John Palfrey and Urs Gasser, *The Connected Parent: An Expert Guide to Parenting in a Digital World* (New York: Hachette, 2020), https://www.hachettebookgroup.com/titles/john-palfrey/the-connected-parent/9781541618022/?lens=basic-books.

24. Howard Gardner, *Multiple Intelligences: The Theory in Practice* (New York: Basic Books, 1993).

25. As quoted in R. Bruce Rich, "Professor Howard Gardner Discusses His Memoir, A Synthesizing Mind," ALI Social Impact Review, September 30, 2023, https://www.sir.advancedleadership.harvard.edu/articles/professor-howard-gardner-discusses-his-memoir-a-synthesizing-mind.

26. Vaidhyanathan, "Generational Myth."

27. Avery Forman, "When Working Harder Doesn't Work, Time to Reinvent Your Career," HBS Working Knowledge, February 15, 2022, http://hbswk.hbs.edu/item/when-working-harder-doesnt-work-time-to-reinvent-your-career.

28. Richard Powers, *The Overstory: A Novel* (New York: W. W. Norton, 2018).

29. Brittne Kakulla, "2023 Tech Trends: No End in Sight for Age 50+ Market Growth," AARP, 2023, https://doi.org/10.26419/res.00584.001; Brittne Kakulla, *Home Tech 2020 AARP: Amerispeak Project and AAPOR Transparency Initiative Report* (Washington, DC: AARP Research, June 28, 2021), https://doi.org/10.26419/res.00420.015; Faverio, "Share of Those 65 and Older"; Pew Research Center, *Older Adults and Technology Use* (Washington, DC: Pew Research Center, 2014), http://www.pewinternet.org/2014/04/03/older-adults-and-technology-use/.

CHAPTER 2

1. Jonathan Gruber, Eszter Hargittai, and Minh Hao Nguyen, "The Value of Face-to-Face Communication in the Digital World: What People Miss about in-Person Interactions When Those Are Limited," *Studies in Communication Sciences* 22 (October 27, 2022): 417–35, https://doi.org/10.24434/j.scoms.2022.03.3340.

2. Minh Hao Nguyen et al., "Trading Spaces: How and Why Older Adults Disconnect from and Switch between Digital Media," *Information Society*, August 23, 2021, 1–13, https://doi.org/10.1080/01972243.2021.1960659.

3. Alessandro Caliandro et al., "Older People and Smartphone Practices in Everyday Life: An Inquire on Digital Sociality of Italian Older Users," *Communication Review* 24, no. 1 (January 2, 2021): 47–78, https://doi.org/10.1080/10714421.2021.1904771; Loredana Ivan and Mireia Fernández-Ardèvol, "Older People

and the Use of ICTs to Communicate with Children and Grandchildren," *Transnational Social Review* 7, no. 1 (January 2, 2017): 41–55, https://doi.org/10.1080/21931674.2016.1277861.

4. Ronald W. Berkowsky, Joseph Sharit, and Sara J. Czaja, "Factors Predicting Decisions About Technology Adoption Among Older Adults," *Innovation in Aging* 1, no. 3 (November 1, 2017), https://doi.org/10.1093/geroni/igy002.

5. Kerryellen G. Vroman, Sajay Arthanat, and Catherine Lysack, "'Who over 65 Is Online?' Older Adults' Dispositions toward Information Communication Technology," *Computers in Human Behavior* 43 (February 1, 2015): 156–66, https://doi.org/10.1016/j.chb.2014.10.018.

6. Anabel Quan-Haase, Kim Martin, and Kathleen Schreurs, "Interviews with Digital Seniors: ICT Use in the Context of Everyday Life," *Information, Communication & Society* 19, no. 5 (2016): 691–707, https://doi.org/10.1080/1369118X.2016.1140217; Anabel Quan-Haase, Guang Ying Mo, and Barry Wellman, "Connected Seniors: How Older Adults in East York Exchange Social Support Online and Offline," *Information, Communication & Society* 20, no. 7 (2017): 967–83, https://doi.org/10.1080/1369118X.2017.1305428; Anabel Quan-Haase, Molly-Gloria Harper, and Alice Hwang, "Digital Media Use and Social Inclusion: A Case Study of East York Older Adults," in *Vulnerable People and Digital Inclusion: Theoretical and Applied Perspectives*, ed. Panayiota Tsatsou (Cham, Switz.: Springer International, 2022), 189–209, https://doi.org/10.1007/978-3-030-94122-2_10; Gemma Wilson et al., "Understanding Older Adults' Use of Social Technology and the Factors Influencing Use," *Ageing & Society* 43, no. 1 (January 2023): 222–45, https://doi.org/10.1017/S0144686X21000490; Shupei Yuan et al., "What Do They Like? Communication Preferences and Patterns of Older Adults in the United States: The Role of Technology," *Educational Gerontology* 42, no. 3 (2016): 163–74, https://doi.org/10.1080/03601277.2015.1083392.

7. Michael T. Bixter et al., "Understanding the Use and Non-Use of Social Communication Technologies by Older Adults: A Qualitative Test and Extension of the UTAUT Model," *Gerontechnology: International Journal on the Fundamental Aspects of Technology to Serve the Ageing Society* 18, no. 2 (2019): 70–88; Caliandro et al., "Older People and Smartphone Practices"; Quan-Haase, Mo, and Wellman, "Connected Seniors."

8. Shelia R. Cotten, William A. Anderson, and Brandi M. McCullough, "Impact of Internet Use on Loneliness and Contact with Others among Older Adults: Cross-Sectional Analysis," *Journal of Medical Internet Research* 15, no. 2 (2013): e39, https://doi.org/10.2196/jmir.2306.

9. Hargittai, *Connected in Isolation*.

10. Sebastiaan T. M. Peek et al., "Older Adults' Reasons for Using Technology While Aging in Place," *Gerontology* 62, no. 2 (2016): 226–37, https://doi.org/10.1159/000430949.

11. Jyoti Choudrie et al., "Investigating the Adoption and Use of Smartphones

in the UK: A Silver-Surfers Perspective" (Hertfordshire Business School Working Paper, University of Hertfordshire, 2014), http://uhra.herts.ac.uk/handle/2299/13507; Amy L. Gonzales, "The Contemporary US Digital Divide: From Initial Access to Technology Maintenance," *Information, Communication & Society* 19, no. 2 (February 1, 2016): 234–48, https://doi.org/10.1080/1369118X.2015.1050438; Alexis Hope, Ted Schwaba, and Anne Marie Piper, "Understanding Digital and Material Social Communications for Older Adults," in *SIGCHI Conference on Human Factors in Computing Systems*, CHI '14 (New York: Association for Computing Machinery, 2014), 3903–12, https://doi.org/10.1145/2556288.2557133.

12. Wilson et al., "Understanding Older Adults' Use."

13. Jošt Bartol et al., "The Roles of Perceived Privacy Control, Internet Privacy Concerns and Internet Skills in the Direct and Indirect Internet Uses of Older Adults: Conceptual Integration and Empirical Testing of a Theoretical Model," *New Media & Society* 26, no. 8 (September 20, 2022): 4490–4510, https://doi.org/10.1177/14614448221122734; Janet Chang, Carolyn McAllister, and Rosemary McCaslin, "Correlates of, and Barriers to, Internet Use among Older Adults," *Journal of Gerontological Social Work* 58, no. 1 (2015): 66–85, https://doi.org/10.1080/01634372.2014.913754; Ceci Diehl et al., "Perceptions on Extending the Use of Technology after the COVID-19 Pandemic Resolves: A Qualitative Study with Older Adults," *International Journal of Environmental Research and Public Health* 19, no. 21 (January 2022): 14152, https://doi.org/10.3390/ijerph192114152; Isioma Elueze and Anabel Quan-Haase, "Privacy Attitudes and Concerns in the Digital Lives of Older Adults: Westin's Privacy Attitude Typology Revisited," *American Behavioral Scientist* 62, no. 10 (September 1, 2018): 1372–91, https://doi.org/10.1177/0002764218787026; Cara Bailey Fausset et al., "Older Adults' Perceptions and Use of Technology: A Novel Approach," in *Universal Access in Human-Computer Interaction: User and Context Diversity*, ed. Constantine Stephanidis and Margherita Antona, Lecture Notes in Computer Science (Berlin: Springer Berlin Heidelberg, 2013), 51–58; Hope, Schwaba, and Piper, "Understanding Digital and Material Social Communications for Older Adults"; Bob Lee, Yiwei Chen, and Lynne Hewitt, "Age Differences in Constraints Encountered by Seniors in Their Use of Computers and the Internet," *Computers in Human Behavior*, Group Awareness in CSCL Environments, 27, no. 3 (May 1, 2011): 1231–37, https://doi.org/10.1016/j.chb.2011.01.003; Marika Lüders and Petter Bae Brandtzæg, "'My Children Tell Me It's so Simple': A Mixed-Methods Approach to Understand Older Non-Users' Perceptions of Social Networking Sites," *New Media & Society* 19, no. 2 (2017): 181–98, https://doi.org/10.1177/1461444814554064; Oluwagbemiga Oyinlola, "Social Media Usage among Older Adults: Insights from Nigeria," *Activities, Adaptation & Aging* 46, no. 4 (October 2, 2022): 343–73, https://doi.org/10.1080/01924788.2022.2044975; Bo Xie et al., "Understanding and Changing Older Adults' Perceptions and Learning of Social Media," *Educational Gerontology* 38, no. 4 (April 1, 2012): 282–96, https://doi.org/10.1080/03601277.2010.544580.

14. Mutong Li, "Factors Influencing Digital Disconnection among the Elderly" (2021 2nd International Conference on Economics, Education and Social Research, UK: Francis Academic Press, 2021), 580–85, https://doi.org/10.25236/iceesr.2021.098; Eleftheria Vaportzis, Maria Giatsi Clausen, and Alan J. Gow, "Older Adults Perceptions of Technology and Barriers to Interacting with Tablet Computers: A Focus Group Study," *Frontiers in Psychology* 8 (2017), https://doi.org/10.3389/fpsyg.2017.01687.

15. Nguyen et al., "Trading Spaces."

16. Ivan and Fernández-Ardèvol, "Older People and the Use of ICTs to Communicate with Children and Grandchildren"; Nguyen et al., "Trading Spaces."

17. John Palfrey and Urs Gasser, *Interop: The Promise and Perils of Highly Interconnected Systems* (New York: Basic Books, 2012).

18. Quan-Haase, Mo, and Wellman, "Connected Seniors"; Pavica Sheldon, Mary Grace Antony, and Lynn Johnson Ware, "Baby Boomers' Use of Facebook and Instagram: Uses and Gratifications Theory and Contextual Age Indicators," *Heliyon* 7, no. 4 (April 1, 2021): e06670, https://doi.org/10.1016/j.heliyon.2021.e06670.

19. Caroline Bell et al., "Examining Social Media Use among Older Adults," in *HT '13: Proceedings of the 24th ACM Conference on Hypertext and Social Media* (New York: Association for Computing Machinery, 2013), 158–63, https://doi.org/10.1145/2481492.2481509; Alexander J. A. M. van Deursen, Jan A. G. M. van Dijk, and Peter M. ten Klooster, "Increasing Inequalities in What We Do Online: A Longitudinal Cross Sectional Analysis of Internet Activities among the Dutch Population (2010 to 2013) over Gender, Age, Education, and Income," *Telematics and Informatics* 32, no. 2 (2015): 259–72, https://doi.org/10.1016/j.tele.2014.09.003.

20. Thomas N. Friemel, "The Digital Divide Has Grown Old: Determinants of a Digital Divide among Seniors," *New Media & Society* 18, no. 2 (2016): 313–31, https://doi.org/10.1177/1461444814538648; Eszter Hargittai and Kerry Dobransky, "Old Dogs, New Clicks: Digital Inequality in Skills and Uses among Older Adults," *Canadian Journal of Communication* 42, no. 2 (2017): 195–212, https://doi.org/10.22230/cjc2017v42n2a3176.

21. Emanuela Sala, Alessandra Gaia, and Gabriele Cerati, "The Gray Digital Divide in Social Networking Site Use in Europe: Results from a Quantitative Study," *Social Science Computer Review* 40, no. 2 (April 1, 2022): 328–45, https://doi.org/10.1177/0894439320909507; Vroman, Arthanat, and Lysack, "Who over 65 Is Online?"; Yuan et al., "What Do They Like?"

22. Francisco Javier Rondán-Cataluña et al., "Social Network Communications in Chilean Older Adults," *International Journal of Environmental Research and Public Health* 17, no. 17 (January 2020): 6078, https://doi.org/10.3390/ijerph17176078.

23. Eszter Hargittai, "Is Bigger Always Better? Potential Biases of Big Data Derived from Social Network Sites," *Annals of the American Academy of*

Political and Social Science 659, no. 1 (2015): 63–76, https://doi.org/10.1177/0002716215570866.

24. Melissa De Regge et al., "Personal and Interpersonal Drivers That Contribute to the Intention to Use Gerontechnologies," *Gerontology* 66, no. 2 (2020): 176–86, https://doi.org/10.1159/000502113.

25. Hsin-yi Sandy Tsai et al., "Getting Grandma Online: Are Tablets the Answer for Increasing Digital Inclusion for Older Adults in the U.S.?," *Educational Gerontology* 41, no. 10 (October 3, 2015): 695–709, https://doi.org/10.1080/03601277.2015.1048165.

26. Kiran Kappeler, Noemi Festic, and Michael Latzer, "Left Behind in the Digital Society—Growing Social Stratification of Internet Non-Use in Switzerland," in *Media Literacy*, ed. Guido Keel and Wibke Weber (Baden-Baden, Ger.: Nomos, 2021), 207–24, https://doi.org/10.5771/9783748920656.

27. Arif Perdana and Intan Azura Mokhtar, "Seniors' Adoption of Digital Devices and Virtual Event Platforms in Singapore during Covid-19," *Technology in Society* 68 (February 1, 2022): 101817, https://doi.org/10.1016/j.techsoc.2021.101817; Kathleen Schreurs, Anabel Quan-Haase, and Kim Martin, "Problematizing the Digital Literacy Paradox in the Context of Older Adults' ICT Use: Aging, Media Discourse, and Self-Determination," *Canadian Journal of Communication* 42, no. 2 (2017): 359–77, https://doi.org/10.22230/cjc.2017v42n2a3130.

28. R. V. Rikard, Ronald W. Berkowsky, and Shelia R. Cotten, "Discontinued Information and Communication Technology Usage among Older Adults in Continuing Care Retirement Communities in the United States," *Gerontology* 64, no. 2 (2018): 188–200, https://doi.org/10.1159/000482017.

29. Rowena Hill, Lucy R. Betts, and Sarah E. Gardner, "Older Adults' Experiences and Perceptions of Digital Technology: (Dis)Empowerment, Wellbeing, and Inclusion," *Computers in Human Behavior* 48 (July 1, 2015): 415–23, https://doi.org/10.1016/j.chb.2015.01.062; Mohsen Javdan, Maryam Ghasemaghaei, and Mohamed Abouzahra, "Psychological Barriers of Using Wearable Devices by Seniors: A Mixed-Methods Study," *Computers in Human Behavior* 141 (April 1, 2023), https://doi.org/10.1016/j.chb.2022.107615.

30. Rondán-Cataluña et al., "Social Network Communications."

31. Michael Segalov, "'We Don't Hold Anything Back': Meet the Old Gays, TikTok's Most Influential Pensioners," *Observer*, November 19, 2023, sec. Technology, https://www.theguardian.com/technology/2023/nov/19/meet-the-old-gays-tiktok-influencers-pensioners.

32. Reuben Ng and Nicole Indran, "Not Too Old for TikTok: How Older Adults Are Reframing Aging," *Gerontologist* 62, no. 8 (October 1, 2022): 1207–16, https://doi.org/10.1093/geront/gnac055.

33. Ivelina Bibeva, "An Exploration of Older Adults' Motivations for Creating Content on TikTok and the Role This Plays for Fostering New Social Connections"

(master's thesis, Malmö University, 2021), http://urn.kb.se/resolve?urn=urn:nbn:se:mau:diva-46219.

CHAPTER 3

1. Vaidhyanathan, "Generational Myth."

2. danah boyd and Eszter Hargittai, "Facebook Privacy Settings: Who Cares?," *First Monday* 15, no. 8 (2010), http://webuse.org/p/a32/; Teresa Correa, "Digital Skills and Social Media Use: How Internet Skills Are Related to Different Types of Facebook Use among 'Digital Natives,'" *Information, Communication & Society* 19, no. 8 (2016): 1095–1107, https://doi.org/10.1080/1369118X.2015.1084023; Hargittai, "Digital Na(t)Ives?"; Hargittai and Hinnant, "Digital Inequality"; Ellen Johanna Helsper and Rebecca Eynon, "Digital Natives: Where Is the Evidence?," *British Educational Research Journal* 36, no. 3 (2010): 503–20, https://doi.org/10.1080/01411920902989227; Sonia Livingstone, Magdalena Bober, and Ellen Helsper, *Internet Literacy among Children and Young People: Findings from the UK Children Go Online Project* (London: UK Children Go Online, 2005), http://personal.lse.ac.uk/bober/UKCGOonlineLiteracy.pdf.

3. Viivi Korpela, Laura Pajula, and Riitta Hänninen, "Older Adults Learning Digital Skills Together: Peer Tutors' Perspectives on Non-Formal Digital Support," *Media and Communication* 11, no. 3 (2023): 53–62, https://doi.org/10.17645/mac.v11i3.6742; Amanda Hunsaker et al., "Unsung Helpers: Older Adults as a Source of Digital Media Support for Their Peers," *Communication Review*, October 12, 2020, 1–22, https://doi.org/10.1080/10714421.2020.1829307; Paul P. Freddolino et al., "To Help and to Learn: An Exploratory Study of Peer Tutors Teaching Older Adults about Technology," *Journal of Technology in Human Services* 28, no. 4 (2010): 217–39, https://doi.org/10.1080/15228835.2011.565458.

4. Anderson and Perrin, "Tech Adoption Climbs."

5. Lan An et al., "Understanding Confidence of Older Adults for Embracing Mobile Technologies," in *OzCHI '22: Proceedings of the 34th Australian Conference on Human-Computer Interaction* (New York: Association for Computing Machinery, 2023), 38–50, https://doi.org/10.1145/3572921.3576202.

6. Quote Investigator, "They May Forget What You Said, But They Will Never Forget How You Made Them Feel," April 6, 2014, https://quoteinvestigator.com/2014/04/06/they-feel/.

7. Johanna L. H. Birkland, *Gerontechnology: Understanding Older Adult Information and Communication Technology* (Leeds, UK: Emerald, 2019), https://doi.org/10.1108/978-1-78743-291-820191002; Jessica Francis et al., "Aging in the Digital Age: Conceptualizing Technology Adoption and Digital Inequalities," in *Ageing and Digital Technology: Designing and Evaluating Emerging Technologies for Older Adults*, ed. Barbara Barbosa Neves and Frank Vetere (Singapore:

Springer Singapore, 2019), 35–49, https://doi.org/10.1007/978-981-13-3693-5_3; Hunsaker et al., "Unsung Helpers."

8. Friemel, "Digital Divide Has Grown Old."

9. Hunsaker et al., "Unsung Helpers"; Will Marler and Eszter Hargittai, "Division of Digital Labor: Partner Support for Technology Use among Older Adults," *New Media & Society* 26, no. 2 (January 27, 2022): 978–94, https://doi.org/10.1177/14614448211068437.

10. Marler and Hargittai, "Division of Digital Labor."

11. Vesna Dolničar et al., "The Role of Social Support Networks in Proxy Internet Use from the Intergenerational Solidarity Perspective," *Telematics and Informatics* 35, no. 2 (May 1, 2018): 305–17, https://doi.org/10.1016/j.tele.2017.12.005; Neil Selwyn et al., "Going Online on Behalf of Others: An Investigation of 'Proxy' Internet Consumers," Australian Communications Consumer Action Network, 2016, https://accan.org.au/index.php; Gemma Webster and Frances Ryan, "Social Media by Proxy: How Older Adults Work within Their 'Social Networks' to Engage with Social Media," *Information Research* 28, no. 1 (2023): 50–77, https://doi.org/10.47989/irpaper952.

12. Hsin-yi Sandy Tsai, Ruth Shillair, and Shelia R. Cotten, "Social Support and 'Playing Around': An Examination of How Older Adults Acquire Digital Literacy with Tablet Computers," *Journal of Applied Gerontology* 36, no. 1 (January 1, 2017): 29–55, https://doi.org/10.1177/0733464815609440.

13. Amanda Hunsaker et al., "'He Explained It to Me and I Also Did It Myself': How Older Adults Get Support with Their Technology Uses," *Socius* 5 (December 4, 2019): 1–13, https://doi.org/10.1177/2378023119887866.

14. Hunsaker et al., 8.

15. Hyunjin Seo et al., "Evidence-Based Digital Literacy Class for Older, Low-Income African-American Adults," *Journal of Applied Communication Research* 47, no. 2 (March 11, 2019): 130–52, https://doi.org/10.1080/00909882.2019.1587176.

16. Stacy Jo Dixon, "Facebook: Quarterly Number of MAU (Monthly Active Users) Worldwide 2008–2023," Statistica, 2024, https://www.statista.com/statistics/264810/number-of-monthly-active-facebook-users-worldwide/.

17. Eszter Hargittai and Marina Micheli, "Internet Skills and Why They Matter," in *Society and the Internet: How Networks of Information and Communication Are Changing Our Lives*, 2nd ed., ed. Mark Graham and William H. Dutton (Oxford: Oxford University Press, 2019), 109–26.

18. Friemel, "Digital Divide Has Grown Old"; Eszter Hargittai, Anne Marie Piper, and Meredith Ringel Morris, "From Internet Access to Internet Skills: Digital Inequality among Older Adults," *Universal Access in the Information Society* 18, no. 4 (May 3, 2018): 881–90, https://doi.org/10.1007/s10209-018-0617-5.

19. Cédric Courtois and Pieter Verdegem, "With a Little Help from My Friends: An Analysis of the Role of Social Support in Digital Inequalities," *New Media & Society* 18, no. 8 (2014): 1508–27, https://doi.org/10.1177/1461444814562162; Marina

Micheli, Elissa M. Redmiles, and Eszter Hargittai, "Help Wanted: Young Adults' Sources of Support for Questions about Digital Media," *Information, Communication & Society* 23, no. 11 (September 18, 2020): 1655–72, https://doi.org/10.1080/1369118X.2019.1602666; Elissa M. Redmiles, Sean Kross, and Michelle L. Mazurek, "How I Learned to Be Secure: A Census-Representative Survey of Security Advice Sources and Behavior," in *CCS'16: Proceedings of the 2016 ACM SIGSAC Conference on Computer and Communications Security* (Vienna: Association for Computing Machinery, 2016), 666–77, https://doi.org/10.1145/2976749.2978307; Wei Zhou, Takami Yasuda, and Shigeki Yokoi, "Supporting Senior Citizens Using the Internet in China," *Research and Practice in Technology Enhanced Learning* 2, no. 1 (March 1, 2007): 75–101, https://doi.org/10.1142/S1793206807000269.

20. DiMaggio et al., "Digital Inequality"; Eszter Hargittai, "Internet Access and Use in Context," *New Media and Society* 6, no. 1 (2004): 137–43, https://doi.org/10.1177/1461444804042310.

21. Hyunjin Seo, "Community-Based Intervention Research Strategies: Digital Inclusion for Marginalized Populations," in *Research Exposed*, ed. Eszter Hargittai (New York: Columbia University Press, 2021), 245–64, https://doi.org/10.7312/harg18876-013; Seo et al., "Evidence-Based Digital Literacy Class."

CHAPTER 4

1. Federal Bureau of Investigation, *Elder Fraud Report 2021* (Criminal Investigative Division, 2021), https://www.ic3.gov/Media/PDF/AnnualReport/2021_IC3ElderFraudReport.pdf.

2. Federal Bureau of Investigation, "Willie Sutton," FBI, May 9, 2022, https://www.fbi.gov/history/famous-cases/willie-sutton.

3. AARP, *The Longevity Economy: How People over 50 Are Driving Economic and Social Value in the US* (Washington, DC: AARP, 2016), https://www.aarp.org/content/dam/aarp/home-and-family/personal-technology/2016/09/2016-Longevity-Economy-AARP.pdf.

4. FBI, *Elder Fraud Report 2021*.

5. Emily Schmall, "Retirees Are Losing Their Life Savings to Romance Scams. Here's What to Know," *New York Times*, February 3, 2023, sec. Business, https://www.nytimes.com/2023/02/03/business/retiree-romance-scams.html.

6. *Last Week Tonight with John Oliver*, season 11, episode 2, "Pig Butchering Scams," aired February 25, 2024, on HBO, available on YouTube, https://www.youtube.com/watch?v=pLPpl2ISKTg.

7. Eldad Bar Lev, Liviu-George Maha, and Stefan-Catalin Topliceanu, "Financial Frauds' Victim Profiles in Developing Countries," *Frontiers in Psychology* 13 (2022), https://www.frontiersin.org/articles/10.3389/fpsyg.2022.999053; Mark Button, Chris Lewis, and Jacki Tapley, *Fraud Typologies and the Victims of Fraud: Literature Review* (London: National Fraud Authority, 2009); Carlos

Carcach, Adam Craycar, and Glenn Muscat, "The Victimisation of Older Australians," Australian Institute of Criminology, June 4, 2001, https://www.aic.gov.au/publications/tandi/tandi212; Michael Ross, Igor Grossmann, and Emily Schryer, "Contrary to Psychological and Popular Opinion, There Is No Compelling Evidence That Older Adults Are Disproportionately Victimized by Consumer Fraud," *Perspectives on Psychological Science* 9, no. 4 (July 1, 2014): 427–42, https://doi.org/10.1177/1745691614535935.

8. Google, "Making You Safer with 2SV," *Keyword*, February 8, 2022, https://blog.google/technology/safety-security/reducing-account-hijacking/.

9. Maximillian Golla et al., "Driving 2FA Adoption at Scale: Optimizing Two-Factor Authentication Notification Design Patterns" (paper presented at the 30th USENIX Security Symposium, August 11–13, 2021), https://www.usenix.org/system/files/sec21-golla.pdf.

10. Sumit Agarwal et al., "The Age of Reason: Financial Decisions over the Life-Cycle with Implications for Regulation," *SSRN Electronic Journal*, 2009, https://doi.org/10.2139/ssrn.973790.

11. Agarwal et al.

12. Helena M. Mentis, Galina Madjaroff, and Aaron K. Massey, "Upside and Downside Risk in Online Security for Older Adults with Mild Cognitive Impairment," in *CHI '19: Proceedings of the 2019 CHI Conference on Human Factors in Computing Systems* (Glasgow: Association for Computing Machinery, 2019), 1–13, https://doi.org/10.1145/3290605.3300573.

13. Rajiv Garg and Rahul Telang, "To Be or Not To Be Linked: Job Search Using Online Social Networks by Unemployed Workforce," SSRN Scholarly Paper (Rochester, NY: Social Science Research Network, April 18, 2011), https://papers.ssrn.com/abstract=1813532; Galen A. Grimes et al., "Older Adults' Knowledge of Internet Hazards," *Educational Gerontology* 36, no. 3 (February 11, 2010): 173–92, https://doi.org/10.1080/03601270903183065.

14. Elissa M. Redmiles, Sean Kross, and Michelle L. Mazurek, *Where Is the Digital Divide? Examining the Impact of Socioeconomics on Security and Privacy Outcomes* (technical report, Digital Repository of University of Maryland, 2016), http://drum.lib.umd.edu/bitstream/handle/1903/18867/CS-TR-5050.pdf?sequence=1&isAllowed=y.

15. Heather Chen and Kathleen Magramo, "Finance Worker Pays Out $25 Million after Video Call with Deepfake 'Chief Financial Officer,'" CNN, February 4, 2024, https://www.cnn.com/2024/02/04/asia/deepfake-cfo-scam-hong-kong-intl-hnk/index.html.

16. Jan Bailey et al., "Older Adults and 'Scams': Evidence from the Mass Observation Archive," *The Journal of Adult Protection* 23, no. 1 (January 1, 2021): 57–69, https://doi.org/10.1108/JAP-07-2020-0030.

17. Alisa Frik et al., "Privacy and Security Threat Models and Mitigation Strate-

gies of Older Adults," 2019, 21–40, https://www.usenix.org/conference/soups2019/presentation/frik.

18. Daniela Oliveira et al., "Dissecting Spear Phishing Emails for Older vs Young Adults: On the Interplay of Weapons of Influence and Life Domains in Predicting Susceptibility to Phishing," in *CHI '17: Proceedings of the 2017 CHI Conference on Human Factors in Computing Systems*, 6412–24 (Denver: Association for Computing Machinery, 2017), https://doi.org/10.1145/3025453.3025831.

19. Golla et al., "Driving 2FA Adoption at Scale: Optimizing Two-Factor Authentication Notification Design Patterns."

20. Frik et al., "Privacy and Security Threat Models and Mitigation Strategies of Older Adults."

21. Victoria Kisekka et al., "Investigating Factors Influencing Web-Browsing Safety Efficacy (WSE) Among Older Adults," *Journal of Information Privacy and Security* 11, no. 3 (July 3, 2015): 158–73, https://doi.org/10.1080/15536548.2015.1073534.

22. Sajay Arthanat, "Promoting Information Communication Technology Adoption and Acceptance for Aging-in-Place: A Randomized Controlled Trial," *Journal of Applied Gerontology* 40, no. 5 (November 2019): 471–80, https://doi.org/10.1177/0733464819891045; Sajay Arthanat, Kerryellen G. Vroman, and Catherine Lysack, "A Home-Based Individualized Information Communication Technology Training Program for Older Adults: A Demonstration of Effectiveness and Value," *Disability and Rehabilitation: Assistive Technology* 11, no. 4 (May 18, 2016): 316–24, https://doi.org/10.3109/17483107.2014.974219; Frik et al., "Privacy and Security Threat Models and Mitigation Strategies of Older Adults"; Jeffrey M. Stanton et al., "Analysis of End User Security Behaviors," *Computers & Security* 24, no. 2 (2005): 124–33, https://doi.org/10.1016/j.cose.2004.07.001.

23. Zhiyuan Wan et al., "AppMoD: Helping Older Adults Manage Mobile Security with Online Social Help," *Proceedings of the ACM on Interactive, Mobile, Wearable and Ubiquitous Technologies* 3, no. 4 (December 11, 2019): 1–22, https://doi.org/10.1145/3369819.

24. Hirak Ray et al., "Why Older Adults (Don't) Use Password Managers" (paper presented at the 30th USENIX Security Symposium, August 11–13, 2021), 73–90, https://www.usenix.org/conference/usenixsecurity21/presentation/ray.

25. Arthanat, Vroman, and Lysack, "Home-Based Individualized Information Communication Technology Training Program for Older Adults."

26. Tamir Mendel and Eran Toch, "My Mom Was Getting This Popup," *Proceedings of the ACM on Interactive, Mobile, Wearable and Ubiquitous Technologies* 3, no. 4 (December 11, 2019): 1–20, https://dl.acm.org/doi/abs/10.1145/3369821.

27. Tamir Mendel et al., "An Exploratory Study of Social Support Systems to Help Older Adults in Managing Mobile Safety," in *Proceedings of the 23rd International Conference on Mobile Human-Computer Interaction* (Toulouse, Fr.: Asso-

ciation for Computing Machinery, 2021), 1–13, https://doi.org/10.1145/3447526.3472047.

28. Savanthi Murthy et al., "Individually Vulnerable, Collectively Safe: The Security and Privacy Practices of Households with Older Adults," *Proceedings of the ACM on Human-Computer Interaction* 5, no. CSCW1 (April 13, 2021): 1–24, https://doi.org/10.1145/3449212.

29. Tamir Mendel, "Social Help: Developing Methods to Support Older Adults in Mobile Privacy and Security," in *Adjunct Proceedings of the 2019 ACM International Joint Conference on Pervasive and Ubiquitous Computing and Proceedings of the 2019 ACM International Symposium on Wearable Computers* (London: Association for Computing Machinery, 2019), 383–87, https://doi.org/10.1145/3341162.3349311.

30. Jess Kropczynski et al., "Towards Building Community Collective Efficacy for Managing Digital Privacy and Security within Older Adult Communities," *Proceedings of the ACM on Human-Computer Interaction* 4, no. CSCW3 (January 5, 2021): 1–27, https://doi.org/10.1145/3432954.

31. Zelle, "How to Spot a Pay Yourself Scam," Zelle, July 22, 2022, YouTube video, https://www.youtube.com/watch?v=8Yk9i4uL8dQ.

32. "Kitboga," YouTube, accessed April 20, 2024, https://www.youtube.com/@KitbogaShow.

33. Bank of America, "How to Identify a Bank Scam to Keep Your Account Safe," Bank of America, 2024, https://www.bankofamerica.com/security-center/avoid-bank-scams/.

34. Patrick Ware et al., "Using eHealth Technologies: Interests, Preferences, and Concerns of Older Adults," *Interactive Journal of Medical Research* 6, no. 1 (March 23, 2017), https://doi.org/10.2196/ijmr.4447; Laura Zettel-Watson and Dmitry Tsukerman, "Adoption of Online Health Management Tools among Healthy Older Adults: An Exploratory Study," *Health Informatics Journal* 22, no. 2 (June 1, 2016): 171–83, https://doi.org/10.1177/1460458214544047.

35. Jinsook Kim and Jennifer Gray, "Qualitative Evaluation of an Intervention Program for Sustained Internet Use Among Low-Income Older Adults," *Ageing International* 41, no. 3 (September 1, 2016): 240–53, https://doi.org/10.1007/s12126-015-9235-1.

36. Nicol Turner Lee et al., "Why the Federal Government Needs to Step Up Efforts to Close the Rural Broadband Divide," Rural Broadband Equity Project, October 4, 2022, https://www.brookings.edu/research/why-the-federal-government-needs-to-step-up-their-efforts-to-close-the-rural-broadband-divide/.

37. James Nicholson, Lynne Coventry, and Pamela Briggs, "'If It's Important It Will Be a Headline': Cybersecurity Information Seeking in Older Adults," in *CHI '19: Proceedings of the 2019 CHI Conference on Human Factors in Computing Systems*, 1–11 (Glasgow: Association for Computing Machinery, 2019), https://doi.org/10.1145/3290605.3300579.

38. Sanchari Das et al., "Why Don't Older Adults Adopt Two-Factor Authentication?," SSRN Scholarly Paper (Rochester, NY, April 25, 2020), https://papers.ssrn.com/abstract=3577820.

39. Carol Witherell and Nel Noddings, *Stories Lives Tell: Narrative and Dialogue in Education* (New York: Teachers College Press, 1991).

40. Charlotte Cowles, "How I Got Scammed Out of $50,000," *New York Magazine*, February 12, 2024, https://www.thecut.com/article/amazon-scam-call-ftc-arrest-warrants.html.

41. Chen and Magramo, "Finance Worker Pays Out $25 Million."

42. Jasmine Kazlauskas, "Cruel Way Scammers Stole Dad's Life Savings," News.Com.Au, March 17, 2024, sec. Banking, https://www.news.com.au/finance/business/banking/cruel-way-scammers-stole-dads-life-savings/news-story/8438a9b7c07e9fa905d5fecc82dace4a.

43. Thibault Spirlet, "A 74-Year-Old Man Lost $50,000 Life Savings after Downloading an App to Order Peking Duck," Business Insider, 2023, https://www.businessinsider.com/man-lost-50k-lifesavings-downloading-app-for-peking-duck-food-2023-10.

44. Emilee Rader, Rick Wash, and Brandon Brooks, "Stories as Informal Lessons about Security," in *SOUPS '12: Proceedings of the Eighth Symposium on Usable Privacy and Security* (New York: Association for Computing Machinery, 2012), 1–17, https://doi.org/10.1145/2335356.2335364.

45. Mendel et al., "Exploratory Study"; Tamir Mendel et al., "Toward Proactive Support for Older Adults: Predicting the Right Moment for Providing Mobile Safety Help," *Proceedings of the ACM on Interactive, Mobile, Wearable and Ubiquitous Technologies* 6, no. 1 (March 29, 2022): 1–25, https://doi.org/10.1145/3517249.

CHAPTER 5

1. Sami Alkhatib et al., "'Who Wants to Know All This Stuff?!': Understanding Older Adults' Privacy Concerns in Aged Care Monitoring Devices," *Interacting with Computers* 33, no. 5 (September 2021): 481–98, https://doi.org/10.1093/iwc/iwab029; Kiran Kappeler, Noemi Festic, and Michael Latzer, "'A Mix of Paranoia and Rebelliousness': Manifestations, Motives, and Consequences of Resistance to Digital Media," *Media Studies* 17, no. 2 (2023): 125–45, https://doi.org/10.5167/uzh-258476.

2. Elueze and Quan-Haase, "Privacy Attitudes and Concerns in the Digital Lives of Older Adults"; Vikram Mehta et al., "Privacy Care: A Tangible Interaction Framework for Privacy Management," *ACM Transactions on Internet Technology* 21, no. 1 (February 2021): 1–32, https://doi.org/10.1145/3430506; Shengzhi Wang et al., "Technology to Support Aging in Place: Older Adults' Perspectives," *Healthcare* 7, no. 2 (June 2019): 60, https://doi.org/10.3390/healthcare7020060.

3. Thora Knight, Xiaojun Yuan, and DeeDee Bennett Gayle, "Illuminating Pri-

vacy and Security Concerns in Older Adults' Technology Adoption," *Work, Aging and Retirement* 10, no. 1 (January 2024): 57–60, https://doi.org/10.1093/workar/waac032.

4. AARP, *Topical Spotlight: Digital Privacy. Privacy Concerns Remain a Barrier* (Washington, DC: AARP Research, August 2021), https://doi.org/10.26419/res.00420.007; Edward C. Baig, "Older Adults Wary About Their Privacy Online," AARP, April 23, 2021, https://www.aarp.org/home-family/personal-technology/info-2021/companies-address-online-privacy-concerns.html.

5. Anabel Quan-Haase and Isioma Elueze, "Revisiting the Privacy Paradox: Concerns and Protection Strategies in the Social Media Experiences of Older Adults," in *Proceedings of the 9th International Conference on Social Media and Society* (Copenhagen: Association for Computing Machinery, 2018), 150–59, https://doi.org/10.1145/3217804.3217907.

6. Andrew McNeill et al., "Functional Privacy Concerns of Older Adults about Pervasive Health-Monitoring Systems," in *Proceedings of the 10th International Conference on PErvasive Technologies Related to Assistive Environments* (Rhodes, Greece: Association for Computing Machinery, 2017), 96–102, https://doi.org/10.1145/3056540.3056559.

7. Alice Marwick and Eszter Hargittai, "Nothing to Hide, Nothing to Lose? Incentives and Disincentives to Sharing Information with Institutions Online," *Information, Communication & Society*, March 29, 2018, 1–17, https://doi.org/10.1080/1369118X.2018.1450432.

8. Anabel Quan-Haase and Dennis Ho, "Online Privacy Concerns and Privacy Protection Strategies among Older Adults in East York, Canada," *Journal of the Association for Information Science and Technology* 71, no. 9 (2020): 1089–1102, https://doi.org/10.1002/asi.24364.

9. AARP, *Topical Spotlight*; Frik et al., "Privacy and Security Threat Models."

10. Ray et al., "Why Older Adults (Don't) Use Password Managers."

11. AARP, *Privacy, Storage, and Usage: A Look at How Older Adults View Big Data in Health Care* (Washington, DC: AARP Research, April 2021), 5, https://doi.org/10.26419/res.00457.001.

12. Mehta et al., "Privacy Care."

13. Wang et al., "Technology to Support Aging in Place."

14. Quan-Haase and Elueze, "Revisiting the Privacy Paradox."

15. Wang et al., "Technology to Support Aging in Place."

16. AARP, *Topical Spotlight.*

17. Brooke Auxier et al., *Americans and Privacy: Concerned, Confused and Feeling Lack of Control Over Their Personal Information* (Washington, DC: Pew Research Center, 2019), 36, https://www.pewresearch.org/internet/2019/11/15/americans-and-privacy-concerned-confused-and-feeling-lack-of-control-over-their-personal-information/.

18. Sara Bannerman and Angela Orasch, "Privacy and Smart Cities: A Canadian Survey," *Canadian Journal of Urban Research* 29, no. 1 (2020): 17–38.

19. Ray et al., "Why Older Adults (Don't) Use Password Managers."

20. Murat Kezer et al., "Age Differences in Privacy Attitudes, Literacy and Privacy Management on Facebook," *Cyberpsychology: Journal of Psychosocial Research on Cyberspace* 10, no. 1 (May 1, 2016), https://doi.org/10.5817/CP2016-1-2.

21. Casey Fiesler et al., "What (or Who) Is Public? Privacy Settings and Social Media Content Sharing," in *CSCW '17: Proceedings of the 2017 ACM Conference on Computer Supported Cooperative Work and Social Computing* (New York: Association for Computing Machinery, 2017), 567–80, https://doi.org/10.1145/2998181.2998223.

22. Sophie C. Boerman, Sanne Kruikemeier, and Frederik J. Zuiderveen Borgesius, "Exploring Motivations for Online Privacy Protection Behavior: Insights from Panel Data," *Communication Research* 48, no. 7 (October 5, 2018): 931–52, https://doi.org/10.1177/0093650218800915.

23. Karen Albright, "On Unexpected Events: Navigating the Sudden Research Opportunity of 9/11," in *Research Confidential: Solutions to Problems Most Social Scientists Pretend They Never Have*, ed. Eszter Hargittai (Ann Arbor: University of Michigan Press, 2009); Hargittai, *Connected in Isolation*.

24. Barry Brown, *Studying the Internet Experience* (Bristol: Hewlett Packard, 2001).

25. Tobias Dienlin, Philipp K. Masur, and Sabine Trepte, "A Longitudinal Analysis of the Privacy Paradox," *New Media*, n.d., 22.

26. Eszter Hargittai and Alice E. Marwick, "'What Can I Really Do?' Explaining the Privacy Paradox with Online Apathy," *International Journal of Communication* 10, no. 0 (2016): 21; Christoph Lutz, Christian Pieter Hoffmann, and Giulia Ranzini, "Data Capitalism and the User: An Exploration of Privacy Cynicism in Germany," *New Media & Society* 22, no. 7 (July 1, 2020): 1168–87, https://doi.org/10.1177/1461444820912544.

27. Charles Duhigg, "How Companies Learn Your Secrets," *New York Times*, February 16, 2012, https://www.nytimes.com/2012/02/19/magazine/shopping-habits.html.

28. Janet Vertesi, "My Experiment Opting Out of Big Data Made Me Look Like a Criminal," *Time*, May 1, 2014, https://time.com/83200/privacy-internet-big-data-opt-out/.

29. Seo et al., "Evidence-Based Digital Literacy Class."

30. Mehta et al., "Privacy Care," 23.

31. Hunsaker et al., "'He Explained It to Me'"; Hunsaker et al., "Unsung Helpers."

32. Mendel, "Social Help."

33. Frik et al., "Privacy and Security Threat Models."

34. Frik et al.; Quan-Haase and Elueze, "Revisiting the Privacy Paradox."

35. Seo et al., "Evidence-Based Digital Literacy Class."

36. AARP, *Privacy, Storage, and Usage.*

37. Alkhatib et al., "'Who Wants to Know All This Stuff?!'"; Daniel J. Solove, "A Taxonomy of Privacy," *University of Pennsylvania Law Review* 154, no. 3 (January 2006): 477–564, https://doi.org/10.2307/40041279.

38. Gordana Dermody et al., "Factors Influencing Community-Dwelling Older Adults' Readiness to Adopt Smart Home Technology: A Qualitative Exploratory Study," *Journal of Advanced Nursing* 77, no. 12 (2021): 4847–61, https://doi.org/10.1111/jan.14996.

39. Heather Kelly, "For Seniors Using Tech to Age in Place, Surveillance Can Be the Price of Independence," *Washington Post*, November 19, 2021, sec. Your Data and Privacy, https://www.washingtonpost.com/technology/2021/11/19/seniors-smart-home-privacy/.

40. Alkhatib et al., "'Who Wants to Know All This Stuff?!'"

41. Apple Inc., "A Day in the Life of Your Data: A Father-Daughter Day at the Playground," April 2021, https://www.apple.com/privacy/docs/A_Day_in_the_Life_of_Your_Data.pdf.

42. Palfrey and Gasser, *Interop.*

CHAPTER 6

1. Andy Guess and Benjamin A. Lyons, "Misinformation, Disinformation, and Online Propaganda," in *Social Media and Democracy: The State of the Field, Prospects for Reform*, ed. Nathaniel Persily and Joshua A. Tucker (Cambridge: Cambridge University Press, 2020), https://books.google.com/books?id=NEjzDwAAQBAJ&pg=PA10&source=gbs_toc_r&cad=2#v=onepage&q&f=false.

2. HealthDay News, "Older Adults More Likely than Young to Be Fooled by 'Fake News,' Study Says - UPI.Com," UPI, 2022, https://www.upi.com/Science_News/2022/05/07/fake-news-older-adults-more-susceptible/2181651853701/.

3. Alex Hern, "Older People More Likely to Share Fake News on Facebook, Study Finds," *Guardian*, January 10, 2019, sec. Technology, https://www.theguardian.com/technology/2019/jan/10/older-people-more-likely-to-share-fake-news-on-facebook.

4. Nadia Brashier and Daniel Schacter, "Op-Ed: Older People Spread More Fake News, a Deadly Habit in the COVID-19 Pandemic," *Los Angeles Times*, August 7, 2020, https://www.latimes.com/opinion/story/2020-08-07/fake-news-older-people-social-media.

5. Andrew Guess, Jonathan Nagler, and Joshua Tucker, "Less than You Think: Prevalence and Predictors of Fake News Dissemination on Facebook," *Science Advances* 5, no. 1 (January 1, 2019): 1–8, https://doi.org/10.1126/sciadv.aau4586.

6. Hargittai, *Connected in Isolation*; Jon Roozenbeek et al., "Susceptibility to

Misinformation about COVID-19 around the World," *Royal Society Open Science* 7, no. 10 (2020): 201199, https://doi.org/10.1098/rsos.201199; Santosh Vijaykumar et al., "How Shades of Truth and Age Affect Responses to COVID-19 (Mis)Information: Randomized Survey Experiment among WhatsApp Users in UK and Brazil," *Humanities and Social Sciences Communications* 8, no. 1 (March 23, 2021): 1–12, https://doi.org/10.1057/s41599-021-00752-7.

7. Guess, Nagler, and Tucker, "Less than You Think."

8. Xun Wang and Robin A. Cohen, "Health Information Technology Use Among Adults: United States, July-December 2022," NCHS Data Brief, no. 482 (Hyattsville, MD: National Center for Health Statistics, October 31, 2023), https://doi.org/10.15620/cdc:133700.

9. World Health Organization, "Managing the COVID-19 Infodemic: Promoting Healthy Behaviours and Mitigating the Harm from Misinformation and Disinformation," September 23, 2020, https://www.who.int/news/item/23-09-2020-managing-the-covid-19-infodemic-promoting-healthy-behaviours-and-mitigating-the-harm-from-misinformation-and-disinformation.

10. World Health Organization, "Novel Coronavirus (2019-nCoV): Situation Report, 13," February 2, 2020, https://apps.who.int/iris/handle/10665/330778.

11. Liz Hamel et al., "KFF COVID-19 Vaccine Monitor: April 2021," *KFF* (blog), May 6, 2021, https://www.kff.org/coronavirus-covid-19/poll-finding/kff-covid-19-vaccine-monitor-april-2021/.

12. Hargittai, *Connected in Isolation.*

13. Vijaykumar et al., "Shades of Truth."

14. Roozenbeek et al., "Susceptibility to Misinformation about COVID-19."

15. Hargittai, *Connected in Isolation.*

16. Sahil Loomba et al., "Measuring the Impact of COVID-19 Vaccine Misinformation on Vaccination Intent in the UK and USA," *Nature Human Behaviour* 5, no. 3 (March 2021): 337–48, https://doi.org/10.1038/s41562-021-01056-1; Sander van der Linden, "Misinformation: Susceptibility, Spread, and Interventions to Immunize the Public," *Nature Medicine* 28, no. 3 (March 2022): 460–67, https://doi.org/10.1038/s41591-022-01713-6.

17. Soroush Vosoughi, Deb Roy, and Sinan Aral, "The Spread of True and False News Online," *Science* 359, no. 6380 (March 9, 2018): 1146–51, https://doi.org/10.1126/science.aap9559.

18. Nadia Brashier and Daniel Schacter, "Aging in an Era of Fake News," *Current Directions in Psychological Science* 29, no. 3 (June 1, 2020): 316–23, https://doi.org/10.1177/0963721420915872.

19. Jyoti Choudrie et al., "Machine Learning Techniques and Older Adults Processing of Online Information and Misinformation: A Covid 19 Study," *Computers in Human Behavior* 119 (June 1, 2021): 11, https://doi.org/10.1016/j.chb.2021.106716.

20. Ross, Grossmann, and Schryer, "Contrary to Psychological and Popular Opinion."

21. Guess, Nagler, and Tucker, "Less than You Think."

22. Nir Grinberg et al., "Fake News on Twitter during the 2016 U.S. Presidential Election," *Science* 363, no. 6425 (January 25, 2019): 374–78, https://doi.org/10.1126/science.aau2706.

23. Sophie J. Nightingale, Kimberley A. Wade, and Derrick G. Watson, "Investigating Age-Related Differences in Ability to Distinguish between Original and Manipulated Images," *Psychology and Aging* 37, no. 3 (2022): 326–37, https://doi.org/10.1037/pag0000682.

24. Björn Herrmann, "The Perception of Artificial-Intelligence (AI) Based Synthesized Speech in Younger and Older Adults," *International Journal of Speech Technology* 26, no. 2 (July 1, 2023): 395–415, https://doi.org/10.1007/s10772-023-10027-y.

25. Cleusa P. Ferri et al., "Global Prevalence of Dementia: A Delphi Consensus Study," *Lancet* 366, no. 9503 (December 17, 2005): 2112–17, https://doi.org/10.1016/S0140-6736(05)67889-0.

26. Nadia Brashier and Daniel Schacter, "Aging in an Era of Fake News," *Current Directions in Psychological Science* 29, no. 3 (June 1, 2020): 316–23, https://doi.org/10.1177/0963721420915872.

27. Brashier and Schacter; Michael A. DeVito et al., "How People Form Folk Theories of Social Media Feeds and What It Means for How We Study Self-Presentation," in *CHI '18: Proceedings of the 2018 CHI Conference on Human Factors in Computing Systems* (Montreal: Association for Computing Machinery, 2018), 1–12, https://doi.org/10.1145/3173574.3173694; Gordon Pennycook and David G. Rand, "Who Falls for Fake News? The Roles of Bullshit Receptivity, Overclaiming, Familiarity, and Analytic Thinking," *Journal of Personality* 88, no. 2 (2020): 185–200, https://doi.org/10.1111/jopy.12476; Emilee Rader and Rebecca Gray, "Understanding User Beliefs About Algorithmic Curation in the Facebook News Feed," in *Proceedings of the 33rd Annual ACM Conference on Human Factors in Computing Systems* (New York: Association for Computing Machinery, 2015), 173–82, https://doi.org/10.1145/2702123.2702174.

28. Sophie J. Nightingale, Kimberley A. Wade, and Derrick G. Watson, "Can People Identify Original and Manipulated Photos of Real-World Scenes?," *Cognitive Research: Principles and Implications* 2, no. 1 (July 18, 2017): 30, https://doi.org/10.1186/s41235-017-0067-2.

29. Brashier and Schacter, "Aging in an Era of Fake News."

30. Michael J. Poulin and Claudia M. Haase, "Growing to Trust: Evidence That Trust Increases and Sustains Well-Being Across the Life Span," *Social Psychological and Personality Science* 6, no. 6 (August 1, 2015): 614–21, https://doi.org/10.1177/1948550615574301.

31. Imani Munyaka, Eszter Hargittai, and Elissa Redmiles, "The Misinformation Paradox: Older Adults Are Cynical about News Media, but Engage with It

Anyway," *Journal of Online Trust and Safety* 1, no. 4 (September 20, 2022), https://doi.org/10.54501/jots.v1i4.62.

32. Guess, Nagler, and Tucker, "Less than You Think."

33. Munyaka, Hargittai, and Redmiles, "Misinformation Paradox."

34. Brown, "Studying the Internet Experience."

35. Hargittai and Marwick, "'What Can I Really Do?'"; Christian Pieter Hoffmann, Christoph Lutz, and Giulia Ranzini, "Privacy Cynicism: A New Approach to the Privacy Paradox," *Cyberpsychology: Journal of Psychosocial Research on Cyberspace* 10, no. 4 (December 1, 2016), https://doi.org/10.5817/CP2016-4-7.

36. Munyaka, Hargittai, and Redmiles, "Misinformation Paradox."

37. Brashier and Schacter, "Aging in an Era of Fake News."

38. Francesca Tripodi, *The Propagandists' Playbook: How Conservative Elites Manipulate Search and Threaten Democracy* (New Haven, CT: Yale University Press, 2022), https://yalebooks.yale.edu/9780300248944/the-propagandists-playbook.

39. Jon Roozenbeek and Sander van der Linden, "Prebunking: A Psychological 'Vaccine' Against Misinformation," *Journalism & Mass Communication Quarterly* 98, no. 3 (2021): 651–54.

40. Melisa Basol, Jon Roozenbeek, and Sander van der Linden, "Good News about Bad News: Gamified Inoculation Boosts Confidence and Cognitive Immunity Against Fake News," *Journal of Cognition* 3, no. 1 (2020): 2, https://doi.org/10.5334/joc.91.

41. Ryan C. Moore and Jeffrey T. Hancock, "A Digital Media Literacy Intervention for Older Adults Improves Resilience to Fake News," *Scientific Reports* 12, no. 6008 (April 9, 2022): 1–9, https://doi.org/10.1038/s41598-022-08437-0.

CHAPTER 7

1. See Büchi and Hargittai, "Need for Considering Digital Inequality," fig. 1.

2. Centers for Disease Control and Prevention, "Health Risks of Social Isolation and Loneliness," Centers for Disease Control and Prevention, May 8, 2023, https://www.cdc.gov/emotional-wellbeing/social-connectedness/loneliness.htm.

3. Susannah Fox, *Rebel Health* (Cambridge, MA: MIT Press, 2024).

4. Emma Curran et al., "Prevalence and Factors Associated with Anxiety and Depression in Older Adults: Gender Differences in Psychosocial Indicators," *Journal of Affective Disorders* 267 (April 15, 2020): 114–22, https://doi.org/10.1016/j.jad.2020.02.018; Julián Alfredo Fernández-Niño et al., "Work Status, Retirement, and Depression in Older Adults: An Analysis of Six Countries Based on the Study on Global Ageing and Adult Health (SAGE)," *SSM - Population Health* 6 (December 2018): 1–8, https://doi.org/10.1016/j.ssmph.2018.07.008.

5. Robert Waldinger and Marc Schulz, "Essay: The Lifelong Power of Close

Relationships," *Wall Street Journal*, January 13, 2023, sec. Life, https://www.wsj.com/articles/the-lifelong-power-of-close-relationships-11673625450.

6. Agnes Szabo et al., "Longitudinal Analysis of the Relationship Between Purposes of Internet Use and Well-Being Among Older Adults," *Gerontologist* 59, no. 1 (January 9, 2019): 58–68, https://doi.org/10.1093/geront/gny036.

7. Julie Erickson and Genevieve M. Johnson, "Internet Use and Psychological Wellness during Late Adulthood," *Canadian Journal on Aging* 30, no. 2 (2011): 197–209, https://doi.org/10.1017/S0714980811000109; Tamara Sims, Andrew E. Reed, and Dawn C. Carr, "Information and Communication Technology Use Is Related to Higher Well-Being Among the Oldest-Old," *Journals of Gerontology: Series B* 72, no. 5 (September 1, 2017): 761–70, https://doi.org/10.1093/geronb/gbw130; Shima Sum et al., "Internet Use and Loneliness in Older Adults," *CyberPsychology & Behavior* 11, no. 2 (April 2008): 208–11, https://doi.org/10.1089/cpb.2007.0010.

8. Szabo et al., "Longitudinal Analysis," 60.

9. Galit Nimrod, "Aging Well in the Digital Age: Technology in Processes of Selective Optimization with Compensation," ed. Jan Warren-Findlow, *Journals of Gerontology: Series B*, August 27, 2019, gbz111, https://doi.org/10.1093/geronb/gbz111.

10. Szabo et al., "Longitudinal Analysis."

11. Roy F. Baumeister and Mark R. Leary, "The Need to Belong: Desire for Interpersonal Attachments as a Fundamental Human Motivation," *Psychological Bulletin* 117, no. 3 (1995): 497–529, https://doi.org/10.1037/0033-2909.117.3.497.

12. Moira Burke, Cameron Marlow, and Thomas Lento, "Social Network Activity and Social Well-Being," in *CHI '10: Proceedings of the SIGCHI Conference on Human Factors in Computing Systems* (New York: Association for Computing Machinery, 2010), 1909–12, https://doi.org/10.1145/1753326.1753613; Minh Hao Nguyen, Amanda Hunsaker, and Eszter Hargittai, "Older Adults' Online Social Engagement and Social Capital: The Moderating Role of Internet Skills," *Information, Communication & Society*, August 25, 2020, 1–17, https://doi.org/10.1080/1369118X.2020.1804980.

13. Chang, McAllister, and McCaslin, "Internet Use among Older Adults"; Elizabeth M. Seabrook, Margaret L. Kern, and Nikki S. Rickard, "Social Networking Sites, Depression, and Anxiety: A Systematic Review," *JMIR Mental Health* 3, no. 4 (November 23, 2016), https://doi.org/10.2196/mental.5842.

14. Matthias Hofer and Eszter Hargittai, "Online Social Engagement, Depression, and Anxiety among Older Adults," *New Media & Society* 26, no. 1 (November 10, 2021): 113–30, https://doi.org/10.1177/14614448211054377.

15. Rebecca P. Yu, Nicole B. Ellison, and Cliff Lampe, "Facebook Use and Its Role in Shaping Access to Social Benefits among Older Adults," *Journal of Broadcasting & Electronic Media* 62, no. 1 (January 2, 2018): 71–90, https://doi.org/10.1080/08838151.2017.1402905.

16. Yu, Ellison, and Lampe, 78.

17. Anna K. Forsman and Johanna Nordmyr, "Psychosocial Links Between Internet Use and Mental Health in Later Life: A Systematic Review of Quantitative and Qualitative Evidence," *Journal of Applied Gerontology* 36, no. 12 (December 1, 2017): 1471–1518, https://doi.org/10.1177/0733464815595509; Nimrod, "Aging Well in the Digital Age," August 27, 2019.

18. Kerry Hannon, "Eldera: The New Global Intergenerational Mentoring Program," *Forbes*, January 5, 2021, https://www.forbes.com/sites/nextavenue/2021/01/05/eldera-the-new-global-intergenerational-mentoring-program/.

19. World Health Organization, "Mental Disorders," April 9, 2018, http://www.who.int/news-room/fact-sheets/detail/mental-disorders.

20. World Health Organization, "Mental Health of Older Adults," December 12, 2017, http://www.who.int/news-room/fact-sheets/detail/mental-health-of-older-adults.

21. Christina E. Miyawaki, "Association of Social Isolation and Health across Different Racial and Ethnic Groups of Older Americans," *Ageing & Society* 35, no. 10 (November 2015): 2201–28, https://doi.org/10.1017/S0144686X14000890.

22. Amanda Hunsaker, Eszter Hargittai, and Anne Marie Piper, "Online Social Connectedness and Anxiety Among Older Adults," *International Journal of Communication* 14, no. 0 (January 28, 2020): 697–725.

23. Rowe and Kahn, "Successful Aging 2.0."

24. Hunsaker, Hargittai, and Piper, "Online Social Connectedness."

25. A. C. Karpinski and A. Duberstein, "A Description of Facebook Use and Academic Performance among Undergraduate and Graduate Students" (paper presented at the Annual Meeting of the American Educational Research Association, April 13–17, 2009).

26. Josh Pasek, Eian More, and Eszter Hargittai, "Facebook and Academic Performance: Reconciling a Media Sensation with Data," *First Monday* 14, no. 5 (2009), https://doi.org/10.5210/fm.v14i5.2498.

27. Matthias Hofer and Allison Eden, "Successful Aging through Television: Selective and Compensatory Television Use and Well-Being," *Journal of Broadcasting & Electronic Media* 64, no. 2 (May 1, 2020): 131–49, https://doi.org/10.1080/08838151.2020.1721259; Robert Kraut et al., "Internet Paradox: A Social Technology That Reduces Social Involvement and Psychological Well-Being?," *American Psychologist* 53, no. 9 (1998): 1017–31, https://doi.org/10.1037/0003-066X.53.9.1017; Dar Meshi, Shelia R. Cotten, and Andrew R. Bender, "Problematic Social Media Use and Perceived Social Isolation in Older Adults: A Cross-Sectional Study," *Gerontology* 66, no. 2 (2020): 160–68, https://doi.org/10.1159/000502577; Anna Schlomann et al., "Use of Information and Communication Technology (ICT) Devices Among the Oldest-Old: Loneliness, Anomie, and Autonomy," *Innovation in Aging* 4, no. 2 (May 1, 2020), https://doi.org/10.1093/geroni/igz050; Seabrook, Kern, and Rickard, "Social Networking Sites."

28. Szabo et al., "Longitudinal Analysis."

29. Leonard Reinecke et al., "Digital Stress over the Life Span: The Effects of Communication Load and Internet Multitasking on Perceived Stress and Psychological Health Impairments in a German Probability Sample," *Media Psychology* 20, no. 1 (January 2, 2017): 90–115, https://doi.org/10.1080/15213269.2015.1121832.

30. Igor Pantic, "Online Social Networking and Mental Health," *Cyberpsychology, Behavior, and Social Networking* 17, no. 10 (September 5, 2014): 652–57, https://doi.org/10.1089/cyber.2014.0070; Edson C. Tandoc, Patrick Ferrucci, and Margaret Duffy, "Facebook Use, Envy, and Depression among College Students: Is Facebooking Depressing?," *Computers in Human Behavior* 43 (February 2015): 139–46, https://doi.org/10.1016/j.chb.2014.10.053.

31. K. Bessière et al., "Effects of Internet Use and Social Resources on Changes in Depression," *Information, Communication & Society* 11, no. 1 (2008): 47–70; Lauren A. Jelenchick, Jens C. Eickhoff, and Megan A. Moreno, "'Facebook Depression?' Social Networking Site Use and Depression in Older Adolescents," *Journal of Adolescent Health* 52, no. 1 (2013): 128–30, https://doi.org/10.1016/j.jadohealth.2012.05.008.

32. Ariz Amoroso Guzman et al., "Social Media Use and Depression in Older Adults: A Systematic Review," *Research in Gerontological Nursing* 16, no. 2 (March 2023): 97–104, https://doi.org/10.3928/19404921-20230220-05.

33. S. Aarts, S. T. M. Peek, and E. J. M. Wouters, "The Relation between Social Network Site Usage and Loneliness and Mental Health in Community-Dwelling Older Adults," *International Journal of Geriatric Psychiatry* 30, no. 9 (2015): 942–49, https://doi.org/10.1002/gps.4241.

34. Kam Man Lau et al., "Social Media and Mental Health in Democracy Movement in Hong Kong: A Population-Based Study," *Computers in Human Behavior* 64 (November 1, 2016): 656–62, https://doi.org/10.1016/j.chb.2016.07.028.

35. Kevin Munger, "Temporal Validity as Meta-Science," *Research & Politics* 10, no. 3 (July 1, 2023): 20531680231187271, https://doi.org/10.1177/20531680231187271.

36. Bang Hyun Kim and Karen Glanz, "Text Messaging to Motivate Walking in Older African Americans: A Randomized Controlled Trial," *American Journal of Preventive Medicine* 44, no. 1 (January 2013): 71–75, https://doi.org/10.1016/j.amepre.2012.09.050.

37. Andre Matthias Müller, Selina Khoo, and Tony Morris, "Text Messaging for Exercise Promotion in Older Adults from an Upper-Middle-Income Country: Randomized Controlled Trial," *Journal of Medical Internet Research* 18, no. 1 (January 7, 2016): e5, https://doi.org/10.2196/jmir.5235.

38. Tilly A. Gurman, Sara E. Rubin, and Amira A. Roess, "Effectiveness of mHealth Behavior Change Communication Interventions in Developing Countries: A Systematic Review of the Literature," *Journal of Health Communication* 17, no. sup1 (May 2, 2012): 82–104, https://doi.org/10.1080/10810730.2011.649160; Müller, Khoo, and Morris, "Text Messaging for Exercise Promotion."

39. Paul DiMaggio and Eszter Hargittai, "From the 'Digital Divide' to 'Digital Inequality': Studying Internet Use as Penetration Increases" (working paper, Princeton University, Woodrow Wilson School of Public and International Affairs, Center for Arts and Cultural Policy Studies, July 2001), https://culturalpolicy.princeton.edu/sites/culturalpolicy/files/wp15_dimaggio_hargittai.pdf; Hargittai, "Digital Reproduction of Inequality"; Hargittai and Micheli, "Internet Skills"; Erik van Ingen and Uwe Matzat, "Inequality in Mobilizing Online Help after a Negative Life Event: The Role of Education, Digital Skills, and Capital-Enhancing Internet Use," *Information, Communication & Society* 21, no. 4 (April 3, 2018): 481–98, https://doi.org/10.1080/1369118X.2017.1293708; Robinson et al., "Digital Inequalities"; Yumei Zhu et al., "The Relationship between Internet Use and Health among Older Adults in China: The Mediating Role of Social Capital," *Healthcare* 9 (2021): 1–15; Nicol Turner-Lee, *Digitally Invisible: How The Internet Is Creating the New Underclass* (Washington, DC: Brookings Institution Press, n.d.).

40. Seeta Peña Gangadharan, "The Downside of Digital Inclusion: Expectations and Experiences of Privacy and Surveillance among Marginal Internet Users," *New Media & Society* 19, no. 4 (April 1, 2017): 597–615, https://doi.org/10.1177/1461444815614053; Marco Gui and Moritz Büchi, "From Use to Overuse: Digital Inequality in the Age of Communication Abundance," *Social Science Computer Review* 39, no. 1 (February 1, 2021): 3–19, https://doi.org/10.1177/0894439319851163.

41. Büchi and Hargittai, "Need for Considering Digital Inequality."

42. Hunsaker, Hargittai, and Piper, "Online Social Connectedness."

43. *Wall Street Journal* Staff, "Facebook's Documents About Instagram and Teens, Published," *Wall Street Journal*, September 30, 2021, sec. Tech, https://www.wsj.com/articles/facebook-documents-instagram-teens-11632953840.

CHAPTER 8

1. Northwestern University, "Newt Minow Leads His First OLLI Study Group: School of Professional Studies, Northwestern University," March 14, 2022, https://sps.northwestern.edu/stories/news-stories/olli_newt-minow-leads-study-group.php.

2. Nahdatul Akma Ahmad et al., "Effectiveness of Instructional Strategies Designed for Older Adults in Learning Digital Technologies: A Systematic Literature Review," *SN Computer Science* 3, no. 2 (January 12, 2022), https://doi.org/10.1007/s42979-022-01016-0.

3. Sean Kross, Eszter Hargittai, and Elissa M. Redmiles, "Characterizing the Online Learning Landscape: What and How People Learn Online," *Proceedings of the ACM on Human-Computer Interaction* 5, no. CSCW1 (April 22, 2021): 1–19, https://doi.org/10.1145/3449220.

4. United Nations, "World Population Prospects, 2017," 24.

5. Toan Nguyen et al., "Access to Mobile Communications by Older People," *Australasian Journal on Ageing* 34, no. 2 (2015): E7–12, https://doi.org/10.1111/ajag.12149.

6. Hunsaker et al., "Unsung Helpers"; Marler and Hargittai, "Division of Digital Labor."

7. Susan M. Ferreira, Sergio Sayago, and Josep Blat, "Older People's Production and Appropriation of Digital Videos: An Ethnographic Study," *Behaviour & Information Technology* 36, no. 6 (June 3, 2017): 557–74, https://doi.org/10.1080/0144929X.2016.1265150.

8. Quan-Haase, Harper, and Hwang, "Digital Media Use."

9. Hargittai, *Connected in Isolation.*

10. Sydney Page, "100-Year-Old Woman Makes Custom Jackets by Hand and Gives Them Away," *Washington Post,* December 29, 2022, https://www.washingtonpost.com/lifestyle/2022/12/27/nancy-epley-100-jackets-virginia/.

11. Amy S. Bruckman, *Should You Believe Wikipedia? Online Communities and the Construction of Knowledge* (Cambridge: Cambridge University Press, 2022), https://doi.org/10.1017/9781108780704.

12. Kross, Hargittai, and Redmiles, "Characterizing the Online Learning Landscape."

13. Aaron Shaw and Eszter Hargittai, "The Pipeline of Online Participation Inequalities: The Case of Wikipedia Editing," *Journal of Communication* 68, no. 1 (February 1, 2018): 143–68, https://doi.org/10.1093/joc/jqx003; Floor Fiers, Aaron Shaw, and Eszter Hargittai, "Generous Attitudes and Online Participation," *Journal of Quantitative Description: Digital Media* 1 (April 26, 2021), https://doi.org/10.51685/jqd.2021.008.

14. Fiers, Shaw, and Hargittai, "Generous Attitudes and Online Participation."

15. Robin Brewer and Anne Marie Piper, "'Tell It Like It Really Is': A Case of Online Content Creation and Sharing Among Older Adult Bloggers," in *CHI '16: Proceedings of the 2016 CHI Conference on Human Factors in Computing Systems* (San Jose, CA: Association for Computing Machinery, 2016), 5529–42, https://doi.org/10.1145/2858036.2858379; Montserrat Celdrán et al., "Exploring the Benefits of Proactive Participation among Adults and Older People by Writing Blogs," *Journal of Gerontological Social Work* 65, no. 3 (April 3, 2022): 320–36, https://doi.org/10.1080/01634372.2021.1965688.

16. S. Andrew Sheppard et al., "Never Too Old, Cold or Dry to Watch the Sky: A Survival Analysis of Citizen Science Volunteerism," *Proceedings of the ACM on Human-Computer Interaction* 1, no. CSCW (December 6, 2017): 1–21, https://doi.org/10.1145/3134729.

17. Lihua Ma et al., "Factors of Web-Based Learning Competence among Urban Chinese Older Adults: Age Differences," *Educational Gerontology* 48, no. 5 (May 4, 2022): 210–23, https://doi.org/10.1080/03601277.2022.2031577.

18. Ching-Ju Chiu et al., "The Attitudes, Impact, and Learning Needs of Older

Adults Using Apps on Touchscreen Mobile Devices: Results from a Pilot Study," *Computers in Human Behavior* 63 (October 1, 2016): 189–97, https://doi.org/10.1016/j.chb.2016.05.020.

19. Chiu et al.; Fabiana M. Gatti, Eleonora Brivio, and Carlo Galimberti, "'The Future Is Ours Too': A Training Process to Enable the Learning Perception and Increase Self-Efficacy in the Use of Tablets in the Elderly," *Educational Gerontology* 43, no. 4 (April 3, 2017): 209–24, https://doi.org/10.1080/03601277.2017.1279952.

20. Ahmad et al., "Effectiveness of Instructional Strategies."

21. Ryan Ebardo, John Byron Tuazon, and Merlin Teodosia Suarez, "We Learn from Each Other: Informal Learning in a Facebook Community of Older Adults," in *Proceedings of the 28th International Conference on Computers in Education*, vol. 2 (Taoyuan City, Taiwan: Asia-Pacific Society for Computers in Education, 2020), 594–601; Ryan Ebardo and Merlin Teodosia Suarez, "Learning Affordances of a Facebook Community of Older Adults: A Netnographic Investigation during COVID-19," in *Proceedings of the 30th International Conference on Computers in Education*, ed. S. Iyer et al. (Taoyuan City, Taiwan: Asia-Pacific Society for Computers in Education, 2022), 140–48, https://icce2022.apsce.net/uploads/P2_W03_019.pdf; Arlind Reuter, Thomas Scharf, and Jan Smeddinck, "Content Creation in Later Life: Reconsidering Older Adults' Digital Participation and Inclusion," in *Proceedings of the ACM on Human-Computer Interaction* 4, no. CSCW3 (January 5, 2021): 1–23, https://doi.org/10.1145/3434166.

22. Eline A. E. Leen and Frieder R. Lang, "Motivation of Computer Based Learning across Adulthood," *Computers in Human Behavior* 29, no. 3 (May 1, 2013): 975–83, https://doi.org/10.1016/j.chb.2012.12.025.

23. Marjolein den Haan et al., "Creating a Social Learning Environment for and by Older Adults in the Use and Adoption of Smartphone Technology to Age in Place," *Frontiers in Public Health* 9 (2021), https://www.frontiersin.org/articles/10.3389/fpubh.2021.568822; Bora Jin, Junghwan Kim, and Lisa M. Baumgartner, "Informal Learning of Older Adults in Using Mobile Devices: A Review of the Literature," *Adult Education Quarterly* 69, no. 2 (May 1, 2019): 120–41, https://doi.org/10.1177/0741713619834726; Sergio Sayago, Paula Forbes, and Josep Blat, "Older People Becoming Successful ICT Learners Over Time: Challenges and Strategies Through an Ethnographical Lens," *Educational Gerontology* 39, no. 7 (July 1, 2013): 527–44, https://doi.org/10.1080/03601277.2012.703583.

24. Haan et al., "Creating a Social Learning Environment."

25. Christian Elias Vazquez et al., "Individualistic Versus Collaborative Learning in an eHealth Literacy Intervention for Older Adults: Quasi-Experimental Study," *JMIR Aging* 6, no. 1 (February 9, 2023): e41809, https://doi.org/10.2196/41809.

26. Arthanat, Vroman, and Lysack, "Home-Based"; Susan M. Ferreira, Sergio Sayago, and Josep Blat, "Learning in Later Life While Engaging in Cross-Generational Digital Content Creation and Playful Educational Activities," in

Game-Based Learning Across the Lifespan: Cross-Generational and Age-Oriented Topics, ed. Margarida Romero et al., Advances in Game-Based Learning (Cham, Switz.: Springer International, 2017), 115–29, https://doi.org/10.1007/978-3-319-41797-4_8; Nguyen et al., "Access to Mobile Communications"; Tsai, Shillair, and Cotten, "Social Support"; Fan Zhang et al., "Situated Learning through Intergenerational Play between Older Adults and Undergraduates," *International Journal of Educational Technology in Higher Education* 14, no. 1 (July 3, 2017): 16, https://doi.org/10.1186/s41239-017-0055-0.

27. AARP, "Books You Can Read to Better Navigate the Technology Landscape," AARP, 2024, https://www.aarp.org/entertainment/books/bookstore/technology-innovation/.

28. Aline Ollevier et al., "How Can Technology Support Ageing in Place in Healthy Older Adults? A Systematic Review," *Public Health Reviews* 41, no. 1 (November 23, 2020): 26, https://doi.org/10.1186/s40985-020-00143-4.

29. Ma et al., "Factors of Web-Based Learning."

30. Quote Investigator, "Give a Man a Fish, and You Feed Him for a Day. Teach a Man to Fish, and You Feed Him for a Lifetime," August 28, 2015, https://quoteinvestigator.com/2015/08/28/fish/.

31. Reuter, Scharf, and Smeddinck, "Content Creation in Later Life."

32. Shelia R. Cotten et al., "Internet Use and Depression among Retired Older Adults in the United States: A Longitudinal Analysis," *Journals of Gerontology: Series B* 69, no. 5 (2014): 763–71, https://doi.org/10.1093/geronb/gbu018; S. Lawrence-Lightfoot, *The Third Chapter: Passion, Risk, and Adventure in the 25 Years After 50* (New York: Farrar, Straus and Giroux, 2009), https://books.google.ch/books?id=PpqgmAEACAAJ.

CHAPTER 9

1. Jessica Taha, Sara J. Czaja, and Joseph Sharit, "Technology Training for Older Job-Seeking Adults: The Efficacy of a Program Offered through a University-Community Collaboration," *Educational Gerontology* 42, no. 4 (April 2, 2016): 276–87, https://doi.org/10.1080/03601277.2015.1109405.

2. Amanda Hunsaker and Eszter Hargittai, "A Review of Internet Use among Older Adults," *New Media & Society* 20, no. 10 (July 16, 2018): 3937–54, https://doi.org/10.1177/1461444818787348.

METHODOLOGICAL APPENDIX

1. Adam J. Berinsky, Michele F. Margolis, and Michael W. Sances, "Separating the Shirkers from the Workers? Making Sure Respondents Pay Attention on Self-Administered Surveys," *American Journal of Political Science* 58, no. 3 (2014): 739–53, https://doi.org/10.1111/ajps.12081.

2. YouGov, "About Our Panel," 2024, https://today.yougov.com/about/panel.

3. Ed Diener et al., "The Satisfaction with Life Scale," *Journal of Personality Assessment* 49, no. 1 (February 1, 1985): 71–75, https://doi.org/10.1207/s15327752jpa4901_13.

4. Tom W. Smith, "Happiness: Time Trends, Seasonal Variations, Intersurvey Differences, and Other Mysteries," *Social Psychology Quarterly* 42, no. 1 (1979): 18–30, https://doi.org/10.2307/3033870.

5. Richard M. Lee and Steven B. Robbins, "Measuring Belongingness: The Social Connectedness and the Social Assurance Scales," *Journal of Counseling Psychology* 42, no. 2 (1995): 232.

6. J. E. Ware and Barbara Gandek, "Overview of the SF-36 Health Survey and the International Quality of Life Assessment (IQOLA) Project," *Journal of Clinical Epidemiology* 51, no. 11 (November 1, 1998): 903–12, https://doi.org/10.1016/S0895-4356(98)00081-X.

7. Carmel Proctor and Roger Tweed, "Measuring Eudaimonic Well-Being," in *Handbook of Eudaimonic Well-Being*, ed. Joar Vittersø, International Handbooks of Quality-of-Life (Cham, Switz.: Springer International, 2016), 277–94, https://doi.org/10.1007/978-3-319-42445-3_18.

8. Aaron T. Beck et al., "An Inventory for Measuring Clinical Anxiety: Psychometric Properties," *Journal of Consulting and Clinical Psychology* 56, no. 6 (1988): 893–97, https://doi.org/10.1037/0022-006X.56.6.893.

9. Ian McDowell, *Measuring Health: A Guide to Rating Scales and Questionnaires*, 3rd ed. (New York: Oxford University Press, 2006); Lenore Sawyer Radloff, "The CES-D Scale: A Self-Report Depression Scale for Research in the General Population," *Applied Psychological Measurement* 1, no. 3 (June 1, 1977): 385–401, https://doi.org/10.1177/014662167700100306.

BIBLIOGRAPHY

AARP. "Books You Can Read to Better Navigate the Technology Landscape." AARP, 2024. https://www.aarp.org/entertainment/books/bookstore/technology-innovation/.

———. *Privacy, Storage, and Usage: A Look at How Older Adults View Big Data in Health Care.* Washington, DC: AARP Research, April 2021. https://doi.org/10.26419/res.00457.001.

———. *Topical Spotlight: Digital Privacy. Privacy Concerns Remain a Barrier.* Washington, DC: AARP Research, August 2021. https://doi.org/10.26419/res.00420.007.

Aarts, S., S. T. M. Peek, and E. J. M. Wouters. "The Relation between Social Network Site Usage and Loneliness and Mental Health in Community-Dwelling Older Adults." *International Journal of Geriatric Psychiatry* 30, no. 9 (2015): 942–49. https://doi.org/10.1002/gps.4241.

Agarwal, Sumit, John C. Driscoll, Xavier Gabaix, and David I. Laibson. "The Age of Reason: Financial Decisions over the Life-Cycle with Implications for Regulation." *SSRN Electronic Journal*, 2009. https://doi.org/10.2139/ssrn.973790.

Aguilera-Hermida, A. Patricia. "Fighting Ageism through Intergenerational Activities, a Transformative Experience." *Journal of Transformative Learning* 7, no. 2 (November 30, 2020): 6–18.

Ahmad, Nahdatul Akma, Muhammad Fairuz Abd Rauf, Najmi Najiha Mohd Zaid, Azaliza Zainal, Tengku Shahrom Tengku Shahdan, and Fariza Hanis Abdul Razak. "Effectiveness of Instructional Strategies Designed for Older Adults in Learning Digital Technologies: A Systematic Literature Review." *SN Computer Science* 3, no. 2 (January 12, 2022). https://doi.org/10.1007/s42979-022-01016-0.

Albright, Karen. "On Unexpected Events: Navigating the Sudden Research Opportunity of 9/11." In *Research Confidential: Solutions to Problems Most Social*

Scientists Pretend They Never Have, edited by Eszter Hargittai. Ann Arbor: University of Michigan Press, 2009.

Alkhatib, Sami, Ryan Kelly, Jenny Waycott, George Buchanan, Marthie Grobler, and Shuo Wang. "'Who Wants to Know All This Stuff?!': Understanding Older Adults' Privacy Concerns in Aged Care Monitoring Devices." *Interacting with Computers* 33, no. 5 (September 2021): 481–98. https://doi.org/10.1093/iwc/iwab029.

An, Lan, Diego Muñoz, Sonja Pedell, and Leon Sterling. "Understanding Confidence of Older Adults for Embracing Mobile Technologies." In *OzCHI '22: Proceedings of the 34th Australian Conference on Human-Computer Interaction*, 38–50. New York: Association for Computing Machinery, 2023. https://doi.org/10.1145/3572921.3576202.

Anderson, Monica, and Andrew Perrin. *Tech Adoption Climbs Among Older Adults*. Washington, DC: Pew Research Center, 2017. http://www.pewinternet.org/2017/05/17/tech-adoption-climbs-among-older-adults/.

Annele, Urtamo, K. Jyväkorpi Satu, and E. Strandberg Timo. "Definitions of Successful Ageing: A Brief Review of a Multidimensional Concept." *Acta Bio Medica: Atenei Parmensis* 90, no. 2 (2019): 359–63. https://doi.org/10.23750/abm.v90i2.8376.

Apple Inc. "A Day in the Life of Your Data: A Father-Daughter Day at the Playground." April 2021. https://www.apple.com/privacy/docs/A_Day_in_the_Life_of_Your_Data.pdf.

Arthanat, Sajay. "Promoting Information Communication Technology Adoption and Acceptance for Aging-in-Place: A Randomized Controlled Trial." *Journal of Applied Gerontology* 40, no. 5 (November 2019): 471–80. https://doi.org/10.1177/0733464819891045.

Arthanat, Sajay, Kerryellen G. Vroman, and Catherine Lysack. "A Home-Based Individualized Information Communication Technology Training Program for Older Adults: A Demonstration of Effectiveness and Value." *Disability and Rehabilitation: Assistive Technology* 11, no. 4 (May 18, 2016): 316–24. https://doi.org/10.3109/17483107.2014.974219.

Auxier, Brooke, Lee Rainie, Monica Anderson, Andrew Perrin, Madhu Kumar and Erica Turner. *Americans and Privacy: Concerned, Confused and Feeling Lack of Control Over Their Personal Information*. Washington, DC: Pew Research Center, 2019. https://www.pewresearch.org/internet/2019/11/15/americans-and-privacy-concerned-confused-and-feeling-lack-of-control-over-their-personal-information/.

Baig, Edward C. "Older Adults Wary About Their Privacy Online." AARP, April 23, 2021. https://www.aarp.org/home-family/personal-technology/info-2021/companies-address-online-privacy-concerns.html.

Bailey, Jan, Louise Taylor, Paul Kingston, and Geoffrey Watts. "Older Adults and 'Scams': Evidence from the Mass Observation Archive." *Journal of Adult Pro-*

tection 23, no. 1 (January 1, 2021): 57–69. https://doi.org/10.1108/JAP-07-2020-0030.

Bank of America. "How to Identify a Bank Scam to Keep Your Account Safe." Bank of America, 2024. https://www.bankofamerica.com/security-center/avoid-bank-scams/.

Bannerman, Sara, and Angela Orasch. "Privacy and Smart Cities: A Canadian Survey." *Canadian Journal of Urban Research* 29, no. 1 (2020): 17–38.

Bar Lev, Eldad, Liviu-George Maha, and Stefan-Catalin Topliceanu. "Financial Frauds' Victim Profiles in Developing Countries." *Frontiers in Psychology* 13 (2022). https://www.frontiersin.org/articles/10.3389/fpsyg.2022.999053.

Barnett, Michael D., and Cassidy M. Adams. "Ageism and Aging Anxiety among Young Adults: Relationships with Contact, Knowledge, Fear of Death, and Optimism." *Educational Gerontology* 44, no. 11 (November 2, 2018): 693–700. https://doi.org/10.1080/03601277.2018.1537163.

Bartol, Jošt, Katja Prevodnik, Vasja Vehovar, and Andraž Petrovčič. "The Roles of Perceived Privacy Control, Internet Privacy Concerns and Internet Skills in the Direct and Indirect Internet Uses of Older Adults: Conceptual Integration and Empirical Testing of a Theoretical Model." *New Media & Society*, 26, no. 8 (September 20, 2022): 4490–4510. https://doi.org/10.1177/14614448221122734.

Basol, Melisa, Jon Roozenbeek, and Sander van der Linden. "Good News about Bad News: Gamified Inoculation Boosts Confidence and Cognitive Immunity Against Fake News." *Journal of Cognition* 3, no.1 (2020): 2. https://doi.org/10.5334/joc.91.

Baumeister, Roy F., and Mark R. Leary. "The Need to Belong: Desire for Interpersonal Attachments as a Fundamental Human Motivation." *Psychological Bulletin* 117, no. 3 (1995): 497–529. https://doi.org/10.1037/0033-2909.117.3.497.

Beck, Aaron T., Norman Epstein, Gary Brown, and Robert A. Steer. "An Inventory for Measuring Clinical Anxiety: Psychometric Properties." *Journal of Consulting and Clinical Psychology* 56, no. 6 (1988): 893–97. https://doi.org/10.1037/0022-006X.56.6.893.

Bell, Caroline, Cara Fausset, Sarah Farmer, Julie Nguyen, Linda Harley, and W. Bradley Fain. "Examining Social Media Use among Older Adults." In *HT '13: Proceedings of the 24th ACM Conference on Hypertext and Social Media*, 158–63. New York: Association for Computing Machinery, 2013. https://doi.org/10.1145/2481492.2481509.

Berinsky, Adam J., Michele F. Margolis, and Michael W. Sances. "Separating the Shirkers from the Workers? Making Sure Respondents Pay Attention on Self-Administered Surveys." *American Journal of Political Science* 58, no. 3 (2014): 739–53. https://doi.org/10.1111/ajps.12081.

Berkowsky, Ronald W., Joseph Sharit, and Sara J. Czaja. "Factors Predicting Decisions About Technology Adoption Among Older Adults." *Innovation in Aging* 1, no. 3 (November 1, 2017). https://doi.org/10.1093/geroni/igy002.

Bessière, K., S. Kiesler, R. Kraut, and B. Boneva. "Effects of Internet Use and Social Resources on Changes in Depression." *Information, Communication & Society* 11, no. 1 (2008): 47–70.

Bibeva, Ivelina. "An Exploration of Older Adults' Motivations for Creating Content on TikTok and the Role This Plays for Fostering New Social Connections." Master's thesis, Malmö University, 2021. http://urn.kb.se/resolve?urn=urn:nbn:se:mau:diva-46219.

Birkland, Johanna L. H. *Gerontechnology: Understanding Older Adult Information and Communication Technology*. Leeds, UK: Emerald, 2019. https://doi.org/10.1108/978-1-78743-291-820191002.

Bixter, Michael T., Kenneth A. Blocker, Tracy L. Mitzner, Akanksha Prakash, and Wendy A. Rogers. "Understanding the Use and Non-Use of Social Communication Technologies by Older Adults: A Qualitative Test and Extension of the UTAUT Model." *Gerontechnology: International Journal on the Fundamental Aspects of Technology to Serve the Ageing Society* 18, no. 2 (2019): 70–88.

Blank, Grant, and Bianca C. Reisdorf. "The Participatory Web." *Information, Communication & Society* 15, no. 4 (2012): 537–54. https://doi.org/10.1080/1369118x.2012.665935.

Boerman, Sophie C., Sanne Kruikemeier, and Frederik J. Zuiderveen Borgesius. "Exploring Motivations for Online Privacy Protection Behavior: Insights from Panel Data." *Communication Research* 48, no. 7 (October 5, 2018): 931–52. https://doi.org/10.1177/0093650218800915.

Bonfadelli, Heinz. "The Internet and Knowledge Gaps: A Theoretical and Empirical Investigation." *European Journal of Communication* 17, no. 1 (2002): 65–84. https://doi.org/10.1177/0267323102017001607.

boyd, danah. *It's Complicated: The Social Lives of Networked Teens*. New Haven, CT: Yale University Press, 2014.

boyd, danah, and Eszter Hargittai. "Facebook Privacy Settings: Who Cares?" *First Monday* 15, no. 8 (2010). http://webuse.org/p/a32/.

Brashier, Nadia, and Daniel Schacter. "Aging in an Era of Fake News." *Current Directions in Psychological Science* 29, no. 3 (June 1, 2020): 316–23. https://doi.org/10.1177/0963721420915872.

———. "Op-Ed: Older People Spread More Fake News, a Deadly Habit in the COVID-19 Pandemic." *Los Angeles Times*, August 7, 2020. https://www.latimes.com/opinion/story/2020-08-07/fake-news-older-people-social-media.

Brewer, Robin, and Anne Marie Piper. "'Tell It Like It Really Is': A Case of Online Content Creation and Sharing Among Older Adult Bloggers." In *CHI '16: Proceedings of the 2016 CHI Conference on Human Factors in Computing Systems*, 5529–42. San Jose, CA: Association for Computing Machinery, 2016. https://doi.org/10.1145/2858036.2858379.

Brown, Barry. *Studying the Internet Experience*. Bristol: Hewlett Packard, 2001.

Bruckman, Amy S. *Should You Believe Wikipedia? Online Communities and the Construction of Knowledge*. Cambridge: Cambridge University Press, 2022. https://doi.org/10.1017/9781108780704.

Büchi, Moritz, and Eszter Hargittai. "A Need for Considering Digital Inequality When Studying Social Media Use and Well-Being." *Social Media + Society* 8, no. 1 (January 1, 2022): 20563051211069125. https://doi.org/10.1177/20563051211069125.

Burke, Moira, Cameron Marlow, and Thomas Lento. "Social Network Activity and Social Well-Being." In *CHI '10: Proceedings of the SIGCHI Conference on Human Factors in Computing Systems*, 1909–12. New York: Association for Computing Machinery, 2010. https://doi.org/10.1145/1753326.1753613.

Burnes, David, Christine Sheppard, Charles R. Henderson, Monica Wassel, Richenda Cope, Chantal Barber, and Karl Pillemer. "Interventions to Reduce Ageism Against Older Adults: A Systematic Review and Meta-Analysis." *American Journal of Public Health* 109, no. 8 (August 2019): 1–9. https://doi.org/10.2105/AJPH.2019.305123.

Button, Mark, Chris Lewis, and Jacki Tapley. *Fraud Typologies and the Victims of Fraud: Literature Review*. London: National Fraud Authority, 2009.

Caliandro, Alessandro, Emma Garavaglia, Valentina Sturiale, and Alice Di Leva. "Older People and Smartphone Practices in Everyday Life: An Inquire on Digital Sociality of Italian Older Users." *Communication Review* 24, no. 1 (January 2, 2021): 47–78. https://doi.org/10.1080/10714421.2021.1904771.

Carcach, Carlos, Adam Craycar, and Glenn Muscat. "The Victimisation of Older Australians." Australian Institute of Criminology, June 4, 2001. https://www.aic.gov.au/publications/tandi/tandi212.

Celdrán, Montserrat, Rodrigo Serrat, Feliciano Villar, and Roger Montserrat. "Exploring the Benefits of Proactive Participation among Adults and Older People by Writing Blogs." *Journal of Gerontological Social Work* 65, no. 3 (April 3, 2022): 320–36. https://doi.org/10.1080/01634372.2021.1965688.

Centers for Disease Control and Prevention. "Health Risks of Social Isolation and Loneliness." Centers for Disease Control and Prevention, May 8, 2023. https://www.cdc.gov/emotional-wellbeing/social-connectedness/loneliness.htm.

Chang, Janet, Carolyn McAllister, and Rosemary McCaslin. "Correlates of, and Barriers to, Internet Use among Older Adults." *Journal of Gerontological Social Work* 58, no. 1 (2015): 66–85. https://doi.org/10.1080/01634372.2014.913754.

Charness, Neil. "A Framework for Choosing Technology Interventions to Promote Successful Longevity: Prevent, Rehabilitate, Augment, Substitute (PRAS)." *Gerontology* 66, no. 2 (2020): 169–75. https://doi.org/10.1159/000502141.

Chen, Heather, and Kathleen Magramo. "Finance Worker Pays Out $25 Million after Video Call with Deepfake 'Chief Financial Officer.'" CNN, February 4, 2024. https://www.cnn.com/2024/02/04/asia/deepfake-cfo-scam-hong-kong-intl-hnk/index.html.

Chiu, Ching-Ju, Yi-Han Hu, Dai-Chan Lin, Fang-Yu Chang, Cheng-Sian Chang, and Cheng-Fung Lai. "The Attitudes, Impact, and Learning Needs of Older Adults Using Apps on Touchscreen Mobile Devices: Results from a Pilot Study." *Computers in Human Behavior* 63 (October 1, 2016): 189–97. https://doi.org/10.1016/j.chb.2016.05.020.

Choudrie, Jyoti, Snehasish Banerjee, Ketan Kotecha, Rahee Walambe, Hema Karende, and Juhi Ameta. "Machine Learning Techniques and Older Adults Processing of Online Information and Misinformation: A Covid 19 Study." *Computers in Human Behavior* 119 (June 1, 2021): 11. https://doi.org/10.1016/j.chb.2021.106716.

Choudrie, Jyoti, Sutee Pheeraphuttharangkoon, Efpraxia Zamani, and George Giaglis. "Investigating the Adoption and Use of Smartphones in the UK: A Silver-Surfers Perspective." Hertfordshire Business School Working Paper, University of Hertfordshire, 2014. http://uhra.herts.ac.uk/handle/2299/13507.

Correa, Teresa. "Digital Skills and Social Media Use: How Internet Skills Are Related to Different Types of Facebook Use among 'Digital Natives.'" *Information, Communication & Society* 19, no. 8 (2016): 1095–1107. https://doi.org/10.1080/1369118X.2015.1084023.

———. "The Participation Divide Among 'Online Experts': Experience, Skills and Psychological Factors as Predictors of College Students' Web Content Creation." *Journal of Computer-Mediated Communication* 16, no. 1 (2010): 71–92. https://doi.org/10.1111/j.1083-6101.2010.01532.x.

Cotten, Shelia R., William A. Anderson, and Brandi M. McCullough. "Impact of Internet Use on Loneliness and Contact with Others among Older Adults: Cross-Sectional Analysis." *Journal of Medical Internet Research* 15, no. 2 (2013): e39. https://doi.org/10.2196/jmir.2306.

Cotten, Shelia R., George Ford, Sherry Ford, and Timothy M. Hale. "Internet Use and Depression among Retired Older Adults in the United States: A Longitudinal Analysis." *Journals of Gerontology: Series B* 69, no. 5 (2014): 763–71. https://doi.org/10.1093/geronb/gbu018.

Courtois, Cédric, and Pieter Verdegem. "With a Little Help from My Friends: An Analysis of the Role of Social Support in Digital Inequalities." *New Media & Society* 18, no. 8 (2014): 1508–27. https://doi.org/10.1177/1461444814562162.

Cowles, Charlotte. "How I Got Scammed Out of $50,000." *New York Magazine*, February 12, 2024. https://www.thecut.com/article/amazon-scam-call-ftc-arrest-warrants.html.

Cremin, Mary Christine. "Feeling Old versus Being Old: Views of Troubled Aging." *Social Science & Medicine* 34, no. 12 (June 1, 1992): 1305–15. https://doi.org/10.1016/0277-9536(92)90139-H.

Curran, Emma, Michael Rosato, Finola Ferry, and Gerard Leavey. "Prevalence and Factors Associated with Anxiety and Depression in Older Adults: Gender Differences in Psychosocial Indicators." *Journal of Affective Disorders* 267 (April 15, 2020): 114–22. https://doi.org/10.1016/j.jad.2020.02.018.

Czaja, Sara J., Walter R. Boot, Neil Charness, and Wendy A. Rogers. *Designing for Older Adults: Principles and Creative Human Factors Approaches*. 3rd ed. Boca Raton, FL: CRC Press, 2019. https://doi.org/10.1201/b22189.

Czaja, Sara J., Chin Chin Lee, Sankaran N. Nair, and Joseph Sharit. "Older Adults and Technology Adoption." *Proceedings of the Human Factors and Ergonomics Society Annual Meeting* 52, no. 2 (September 1, 2008): 139–43. https://doi.org/10.1177/154193120805200201.

Das, Sanchari, Andrew Kim, Ben Jelen, Joshua Streiff, L. Jean Camp, and Lesa Huber. "Why Don't Older Adults Adopt Two-Factor Authentication?" SSRN Scholarly Paper. Rochester, NY, April 25, 2020. https://papers.ssrn.com/abstract=3577820.

De Regge, Melissa, Freek Van Baelen, Gabriela Beirão, Anouk Den Ambtman, Kaat De Pourcq, Joana Carmo Dias, and Jay Kandampully. "Personal and Interpersonal Drivers That Contribute to the Intention to Use Gerontechnologies." *Gerontology* 66, no. 2 (2020): 176–86. https://doi.org/10.1159/000502113.

Dermody, Gordana, Roschelle Fritz, Courtney Glass, Melissa Dunham, and Lisa Whitehead. "Factors Influencing Community-Dwelling Older Adults' Readiness to Adopt Smart Home Technology: A Qualitative Exploratory Study." *Journal of Advanced Nursing* 77, no. 12 (2021): 4847–61. https://doi.org/10.1111/jan.14996.

Deursen, Alexander J. A. M. van, and J. A. G. M. van Dijk. "Internet Skills and the Digital Divide." *New Media & Society* 13, no. 6 (2011): 893–911. https://doi.org/10.1177/1461444810386774.

Deursen, Alexander J. A. M. van, Jan A. G. M. van Dijk, and Peter M. ten Klooster. "Increasing Inequalities in What We Do Online: A Longitudinal Cross Sectional Analysis of Internet Activities among the Dutch Population (2010 to 2013) over Gender, Age, Education, and Income." *Telematics and Informatics* 32, no. 2 (2015): 259–72. https://doi.org/10.1016/j.tele.2014.09.003.

Deursen, Alexander J. A. M. van, Ellen Helsper, Rebecca Eynon, and Jan A. G. M. van Dijk. "The Compoundness and Sequentiality of Digital Inequality." *International Journal of Communication* 11 (2017): 22.

DeVito, Michael A., Jeremy Birnholtz, Jeffery T. Hancock, Megan French, and Sunny Liu. "How People Form Folk Theories of Social Media Feeds and What It Means for How We Study Self-Presentation." In *CHI '18: Proceedings of the 2018 CHI Conference on Human Factors in Computing Systems*, 1–12. Montreal: Association for Computing Machinery, 2018. https://doi.org/10.1145/3173574.3173694.

Diehl, Ceci, Rita Tavares, Taiane Abreu, Ana Margarida Pisco Almeida, Telmo Eduardo Silva, Gonçalo Santinha, Nelson Pacheco Rocha, et al. "Perceptions on Extending the Use of Technology after the COVID-19 Pandemic Resolves: A Qualitative Study with Older Adults." *International Journal of Environmental Research and Public Health* 19, no. 21 (January 2022): 14152. https://doi.org/10.3390/ijerph192114152.

Diener, Ed, Robert A. Emmons, Randy J. Larsen, and Sharon Griffin. "The Satisfaction with Life Scale." *Journal of Personality Assessment* 49, no. 1 (February 1, 1985): 71–75. https://doi.org/10.1207/s15327752jpa4901_13.

Dienlin, Tobias, Philipp K. Masur, and Sabine Trepte. "A Longitudinal Analysis of the Privacy Paradox." *New Media*, n.d., 22.

DiMaggio, Paul, and Eszter Hargittai. "From the 'Digital Divide' to 'Digital Inequality': Studying Internet Use as Penetration Increases." Working paper, Princeton University, Woodrow Wilson School of Public and International Affairs, Center for Arts and Cultural Policy Studies, July 2001. https://culturalpolicy.princeton.edu/sites/culturalpolicy/files/wp15_dimaggio_hargittai.pdf.

DiMaggio, Paul, Eszter Hargittai, Coral Celeste, and Steven Schafer. "Digital Inequality: From Unequal Access to Differentiated Use." In *Social Inequality*, edited by Kathryn Neckerman, 355–400. New York: Russell Sage Foundation, 2004.

Dixon, Stacy Jo. "Facebook: Quarterly Number of MAU (Monthly Active Users) Worldwide 2008-2023." Statistica, 2024. https://www.statista.com/statistics/264810/number-of-monthly-active-facebook-users-worldwide/.

Dobransky, Kerry, and Eszter Hargittai. "The Disability Divide in Internet Access and Use." *Information, Communication and Society* 9, no. 3 (2006): 313–34. https://doi.org/10.1080/13691180600751298.

Dolničar, Vesna, Darja Grošelj, Maša Filipovič Hrast, Vasja Vehovar, and Andraž Petrovčič. "The Role of Social Support Networks in Proxy Internet Use from the Intergenerational Solidarity Perspective." *Telematics and Informatics* 35, no. 2 (May 1, 2018): 305–17. https://doi.org/10.1016/j.tele.2017.12.005.

Drury, Lisbeth, Dominic Abrams, and Hannah J. Swift. *Making Intergenerational Connections—An Evidence Review: What Are They, Why Do They Matter and How to Make More of Them.* London: Age UK, 2017. https://www.ageuk.org.uk/globalassets/age-uk/documents/reports-and-publications/reports-and-briefings/active-communities/rb_2017_making_intergenerational_connections.pdf.

Duhigg, Charles. "How Companies Learn Your Secrets." *New York Times*, February 16, 2012. https://www.nytimes.com/2012/02/19/magazine/shopping-habits.html.

Dutton, William H., and Grant Blank. "The Emergence of Next Generation Internet Users." *International Economics and Economic Policy* 11, no. 1–2 (2014): 29–47. https://doi.org/10.1007/s10368-013-0245-8.

Ebardo, Ryan, and Merlin Teodosia Suarez. "Learning Affordances of a Facebook

Community of Older Adults: A Netnographic Investigation during COVID-19." In *Proceedings of the 30th International Conference on Computers in Education*, edited by S. Iyer et al., 140–48. Taoyuan City, Taiwan: Asia-Pacific Society for Computers in Education, 2022. https://icce2022.apsce.net/uploads/P2_W03_019.pdf.

Ebardo, Ryan, John Byron Tuazon, and Merlin Teodosia Suarez. "We Learn from Each Other: Informal Learning in a Facebook Community of Older Adults." In *Proceedings of the 28th International Conference on Computers in Education*, 594–601. Vol. 2. Taoyuan City, Taiwan: Asia-Pacific Society for Computers in Education, 2020.

Ellison, Nicole B., Charles Steinfield, and Cliff Lampe. "The Benefits of Facebook 'Friends': Social Capital and College Students' Use of Online Social Network Sites." *Journal of Computer-Mediated Communication* 12, no. 4 (2007): 1143–68. https://doi.org/10.1111/j.1083-6101.2007.00367.x.

Elueze, Isioma, and Anabel Quan-Haase. "Privacy Attitudes and Concerns in the Digital Lives of Older Adults: Westin's Privacy Attitude Typology Revisited." *American Behavioral Scientist* 62, no. 10 (September 1, 2018): 1372–91. https://doi.org/10.1177/0002764218787026.

Erickson, Julie, and Genevieve M. Johnson. "Internet Use and Psychological Wellness during Late Adulthood." *Canadian Journal on Aging / La Revue Canadienne Du Vieillissement* 30, no. 2 (2011): 197–209. https://doi.org/10.1017/S0714980811000109.

Eynon, R. "Mapping the Digital Divide in Britain: Implications for Learning and Education." *Learning, Media, and Technology* 34, no. 4 (2009): 277–90. https://doi.org/10.1080/17439880903345874.

Fausset, Cara Bailey, Linda Harley, Sarah Farmer, and Brad Fain. "Older Adults' Perceptions and Use of Technology: A Novel Approach." In *Universal Access in Human-Computer Interaction: User and Context Diversity*, edited by Constantine Stephanidis and Margherita Antona, 51–58. Lecture Notes in Computer Science. Berlin: Springer Berlin Heidelberg, 2013.

Faverio, Michelle. "Share of Those 65 and Older Who Are Tech Users Has Grown in the Past Decade." *Pew Research Center* (blog), January 13, 2022. https://www.pewresearch.org/short-reads/2022/01/13/share-of-those-65-and-older-who-are-tech-users-has-grown-in-the-past-decade/.

Federal Bureau of Investigation. *Elder Fraud Report 2021*. Criminal Investigative Division. https://www.ic3.gov/Media/PDF/AnnualReport/2021_IC3ElderFraudReport.pdf.

———. "Willie Sutton." FBI, May 9, 2022. https://www.fbi.gov/history/famous-cases/willie-sutton.

Fernández-Niño, Julián Alfredo, Laura Juliana Bonilla-Tinoco, Betty Soledad Manrique-Espinoza, Martin Romero-Martínez, and Ana Luisa Sosa-Ortiz. "Work Status, Retirement, and Depression in Older Adults: An Analysis of Six

Countries Based on the Study on Global Ageing and Adult Health (SAGE)." *SSM—Population Health* 6 (December 2018): 1–8. https://doi.org/10.1016/j.ssmph.2018.07.008.

Ferreira, Susan M., Sergio Sayago, and Josep Blat. "Learning in Later Life While Engaging in Cross-Generational Digital Content Creation and Playful Educational Activities." In *Game-Based Learning Across the Lifespan: Cross-Generational and Age-Oriented Topics*, edited by Margarida Romero, Kimberly Sawchuk, Josep Blat, Sergio Sayago, and Hubert Ouellet, 115–29. Advances in Game-Based Learning. Cham, Switz.: Springer International, 2017. https://doi.org/10.1007/978-3-319-41797-4_8.

———. "Older People's Production and Appropriation of Digital Videos: An Ethnographic Study." *Behaviour & Information Technology* 36, no. 6 (June 3, 2017): 557–74. https://doi.org/10.1080/0144929X.2016.1265150.

Ferri, Cleusa P., Martin Prince, Carol Brayne, Henry Brodaty, Laura Fratiglioni, Mary Ganguli, Kathleen Hall, et al. "Global Prevalence of Dementia: A Delphi Consensus Study." *Lancet* 366, no. 9503 (December 17, 2005): 2112–17. https://doi.org/10.1016/S0140-6736(05)67889-0.

Fiers, Floor, Aaron Shaw, and Eszter Hargittai. "Generous Attitudes and Online Participation." *Journal of Quantitative Description: Digital Media* 1 (April 26, 2021). https://doi.org/10.51685/jqd.2021.008.

Fiesler, Casey, Michaelanne Dye, Jessica L. Feuston, Chaya Hiruncharoenvate, C. J. Hutto, Shannon Morrison, Parisa Khanipour Roshan, et al. "What (or Who) Is Public? Privacy Settings and Social Media Content Sharing." In *CSCW '17: Proceedings of the 2017 ACM Conference on Computer Supported Cooperative Work and Social Computing*, 567–80. New York: Association for Computing Machinery, 2017. https://doi.org/10.1145/2998181.2998223.

Forman, Avery. "When Working Harder Doesn't Work, Time to Reinvent Your Career." HBS Working Knowledge, February 15, 2022. http://hbswk.hbs.edu/item/when-working-harder-doesnt-work-time-to-reinvent-your-career.

Forsman, Anna K., and Johanna Nordmyr. "Psychosocial Links Between Internet Use and Mental Health in Later Life: A Systematic Review of Quantitative and Qualitative Evidence." *Journal of Applied Gerontology* 36, no. 12 (December 1, 2017): 1471–1518. https://doi.org/10.1177/0733464815595509.

Fox, Susannah. *Rebel Health*. Cambridge, MA: MIT Press, 2024.

Francis, Jessica, Christopher Ball, Travis Kadylak, and Shelia R. Cotten. "Aging in the Digital Age: Conceptualizing Technology Adoption and Digital Inequalities." In *Ageing and Digital Technology: Designing and Evaluating Emerging Technologies for Older Adults*, edited by Barbara Barbosa Neves and Frank Vetere, 35–49. Singapore: Springer Singapore, 2019. https://doi.org/10.1007/978-981-13-3693-5_3.

Freddolino, Paul P., Vincent W. P. Lee, Chi-Kwong Law, and Cindy Ho. "To Help and to Learn: An Exploratory Study of Peer Tutors Teaching Older Adults about Technology." *Journal of Technology in Human Services* 28, no, 4 (2010): 217–39. https://doi.org/10.1080/15228835.2011.565458.

Freund, Alexandra M., and Michaela Riediger. "Successful Aging." In *Handbook of Psychology*, Vol. 6, *Developmental Psychology*, edited by Richard M. Lerner, M. Ann Easterbrooks, and Jayanthi Mistry, 601–28. 2nd ed. (Hoboken, NJ: John Wiley & Sons, 2003). https://www.wiley.com/en-us/Handbook+of+Psychology%2C+Volume+6%2C+Developmental+Psychology%2C+2nd+Edition-p-9780470768860.

Friemel, Thomas N. "The Digital Divide Has Grown Old: Determinants of a Digital Divide among Seniors." *New Media & Society* 18, no. 2 (2016): 313–31. https://doi.org/10.1177/1461444814538648.

Frik, Alisa, Leysan Nurgalieva, Julia Bernd, Joyce Lee, Florian Schaub, and Serge Egelman. "Privacy and Security Threat Models and Mitigation Strategies of Older Adults," 21–40. Symposium on Usable Privacy and Security, 2019. https://www.usenix.org/conference/soups2019/presentation/frik.

Gangadharan, Seeta Peña. "The Downside of Digital Inclusion: Expectations and Experiences of Privacy and Surveillance among Marginal Internet Users." *New Media & Society* 19, no. 4 (April 1, 2017): 597–615. https://doi.org/10.1177/1461444815614053.

Gardner, Howard. *Multiple Intelligences: The Theory in Practice*. New York: Basic Books, 1993.

Garg, Rajiv, and Rahul Telang. "To Be or Not to Be Linked: Job Search Using Online Social Networks by Unemployed Workforce." SSRN Scholarly Paper. Rochester, NY: Social Science Research Network, April 18, 2011. https://papers.ssrn.com/abstract=1813532.

Gatti, Fabiana M., Eleonora Brivio, and Carlo Galimberti. "'The Future Is Ours Too': A Training Process to Enable the Learning Perception and Increase Self-Efficacy in the Use of Tablets in the Elderly." *Educational Gerontology* 43, no. 4 (April 3, 2017): 209–24. https://doi.org/10.1080/03601277.2017.1279952.

Golla, Maximillian, Grant Ho, Marika Lohmus, Monica Pulluri, and Elissa M. Redmiles. "Driving 2FA Adoption at Scale: Optimizing Two-Factor Authentication Notification Design Patterns." Paper presented at the 30th USENIX Security Symposium, August 11–13, 2021. https://www.usenix.org/system/files/sec21-golla.pdf.

Gonzales, Amy L. "The Contemporary US Digital Divide: From Initial Access to Technology Maintenance." *Information, Communication & Society* 19, no. 2 (February 1, 2016): 234–48. https://doi.org/10.1080/1369118X.2015.1050438.

Google. "Making You Safer with 2SV." *Keyword* (blog), February 8, 2022. https://blog.google/technology/safety-security/reducing-account-hijacking/.

Grimes, Galen A., Michelle G. Hough, Elizabeth Mazur, and Margaret L. Signorella. "Older Adults' Knowledge of Internet Hazards." *Educational Gerontology* 36, no. 3 (February 11, 2010): 173–92. https://doi.org/10.1080/03601270903183065.

Grinberg, Nir, Kenneth Joseph, Lisa Friedland, Briony Swire-Thompson, and David Lazer. "Fake News on Twitter during the 2016 U.S. Presidential Election." *Science* 363, no. 6425 (January 25, 2019): 374–78. https://doi.org/10.1126/science.aau2706.

Gruber, Jonathan, Eszter Hargittai, and Minh Hao Nguyen. "The Value of Face-to-Face Communication in the Digital World: What People Miss about in-Person Interactions When Those Are Limited." *Studies in Communication Sciences* 22 (October 27, 2022): 417–35. https://doi.org/10.24434/j.scoms.2022.03.3340.

Guess, Andrew M., and Benjamin A. Lyons. "Misinformation, Disinformation, and Online Propaganda." In *Social Media and Democracy: The State of the Field, Prospects for Reform*, edited by Nathaniel Persily and Joshua A. Tucker, 10–33. Cambridge: Cambridge University Press, 2020. https://books.google.com/books?id=NEjzDwAAQBAJ&pg=PA10&source=gbs_toc_r&cad=2#v=onepage&q&f=false.

Guess, Andrew, Jonathan Nagler, and Joshua Tucker. "Less than You Think: Prevalence and Predictors of Fake News Dissemination on Facebook." *Science Advances* 5, no. 1 (January 1, 2019): 1–8. https://doi.org/10.1126/sciadv.aau4586.

Gui, Marco, and Moritz Büchi. "From Use to Overuse: Digital Inequality in the Age of Communication Abundance." *Social Science Computer Review* 39, no. 1 (February 1, 2021): 3–19. https://doi.org/10.1177/0894439319851163.

Gurman, Tilly A., Sara E. Rubin, and Amira A. Roess. "Effectiveness of mHealth Behavior Change Communication Interventions in Developing Countries: A Systematic Review of the Literature." *Journal of Health Communication* 17, no. sup1 (May 2, 2012): 82–104. https://doi.org/10.1080/10810730.2011.649160.

Guzman, Ariz Amoroso, Mary-Lynn Brecht, Lynn V. Doering, Paul M. Macey, and Janet C. Mentes. "Social Media Use and Depression in Older Adults: A Systematic Review." *Research in Gerontological Nursing* 16, no. 2 (March 2023): 97–104. https://doi.org/10.3928/19404921-20230220-05.

Haan, Marjolein den, Rens Brankaert, Gail Kenning, and Yuan Lu. "Creating a Social Learning Environment for and by Older Adults in the Use and Adoption of Smartphone Technology to Age in Place." *Frontiers in Public Health* 9 (2021). https://www.frontiersin.org/articles/10.3389/fpubh.2021.568822.

Hamel, Liz, Lunna Lopes, Grace Sparks, Mellisha Stokes, and Mollyann Brodie. "KFF COVID-19 Vaccine Monitor: April 2021." *KFF* (blog), May 6, 2021. https://www.kff.org/coronavirus-covid-19/poll-finding/kff-covid-19-vaccine-monitor-april-2021/.

Hannon, Kerry. "Eldera: The New Global Intergenerational Mentoring Program." *Forbes*, January 5, 2021. https://www.forbes.com/sites/nextavenue/2021/01/05/eldera-the-new-global-intergenerational-mentoring-program/.

Hargittai, Eszter. *Connected in Isolation: Digital Privilege in Unsettled Times*. Cambridge, MA: MIT Press, 2022.

———. "Digital Na(t)Ives? Variation in Internet Skills and Uses among Members of the 'Net Generation'*." *Sociological Inquiry* 80, no. 1 (2010): 92–113. https://doi.org/10.1111/j.1475-682X.2009.00317.x.

———. "The Digital Reproduction of Inequality." In *Social Stratification*, edited by David Grusky, 936–44. Boulder, CO: Westview Press, 2008.

———, ed. *Handbook of Digital Inequality*. Cheltenham, UK: Edward Elgar, 2021.

———. "Internet Access and Use in Context." *New Media and Society* 6, no. 1 (2004): 137–43. https://doi.org/10.1177/1461444804042310.

———. "Is Bigger Always Better? Potential Biases of Big Data Derived from Social Network Sites." *ANNALS of the American Academy of Political and Social Science* 659, no. 1 (2015): 63–76. https://doi.org/10.1177/0002716215570866.

———. "Second-Level Digital Divide: Differences in People's Online Skills." *First Monday* 7, no. 4 (April 1, 2002). http://firstmonday.org/ojs/index.php/fm/article/view/942.

Hargittai, Eszter, and Kerry Dobransky. "Old Dogs, New Clicks: Digital Inequality in Skills and Uses among Older Adults." *Canadian Journal of Communication* 42, no. 2 (2017): 195–212. https://doi.org/10.22230/cjc2017v42n2a3176.

Hargittai, Eszter, and Amanda Hinnant. "Digital Inequality: Differences in Young Adults' Use of the Internet." *Communication Research* 35, no. 5 (2008): 602–21. https://doi.org/10.1177/0093650208321782.

Hargittai, Eszter, and Alice E. Marwick. "'What Can I Really Do?' Explaining the Privacy Paradox with Online Apathy." *International Journal of Communication* 10, no. 0 (2016): 21.

Hargittai, Eszter, and Marina Micheli. "Internet Skills and Why They Matter." In *Society and the Internet. How Networks of Information and Communication Are Changing Our Lives*. 2nd ed. Edited by Mark Graham and William H. Dutton, 109–26. Oxford: Oxford University Press, 2019.

Hargittai, Eszter, Anne Marie Piper, and Meredith Ringel Morris. "From Internet Access to Internet Skills: Digital Inequality among Older Adults." *Universal Access in the Information Society* 18, no. 4 (May 3, 2018): 881–90. https://doi.org/10.1007/s10209-018-0617-5.

Hargittai, Eszter, and G. Walejko. "The Participation Divide: Content Creation and Sharing in the Digital Age." *Information, Communication and Society* 11, no. 2 (2008): 239–56. https://doi.org/10.1080/13691180801946150.

HealthDay News. "Older Adults More Likely than Young to Be Fooled by 'Fake News,' Study Says - UPI.Com." UPI, 2022. https://www.upi.com/Science_News/2022/05/07/fake-news-older-adults-more-susceptible/2181651853701/.

Helsper, Ellen Johanna, and Rebecca Eynon. "Digital Natives: Where Is the Evidence?" *British Educational Research Journal* 36, no. 3 (2010): 503–20. https://doi.org/10.1080/01411920902989227.

Hern, Alex. "Older People More Likely to Share Fake News on Facebook, Study

Finds." *Guardian*, January 10, 2019, sec. Technology. https://www.theguardian.com/technology/2019/jan/10/older-people-more-likely-to-share-fake-news-on-facebook.

Herrmann, Björn. "The Perception of Artificial-Intelligence (AI) Based Synthesized Speech in Younger and Older Adults." *International Journal of Speech Technology* 26, no. 2 (July 1, 2023): 395–415. https://doi.org/10.1007/s10772-023-10027-y.

Hill, Rowena, Lucy R. Betts, and Sarah E. Gardner. "Older Adults' Experiences and Perceptions of Digital Technology: (Dis)Empowerment, Wellbeing, and Inclusion." *Computers in Human Behavior* 48 (July 1, 2015): 415–23. https://doi.org/10.1016/j.chb.2015.01.062.

Hofer, Matthias, and Allison Eden. "Successful Aging through Television: Selective and Compensatory Television Use and Well-Being." *Journal of Broadcasting & Electronic Media* 64, no. 2 (May 1, 2020): 131–49. https://doi.org/10.1080/08838151.2020.1721259.

Hofer, Matthias, and Eszter Hargittai. "Online Social Engagement, Depression, and Anxiety among Older Adults." *New Media & Society* 26, no. 1 (November 10, 2021): 113–30. https://doi.org/10.1177/14614448211054377.

Hoffman, Donna L., Thomas P. Novak, and Ann Schlosser. "The Evolution of the Digital Divide: How Gaps in Internet Access May Impact Electronic Commerce." *Journal of Computer-Mediated Communication* 5, no. 3 (2000). https://doi.org/10.1111/j.1083-6101.2000.tb00341.x.

Hoffmann, Christian Pieter, Christoph Lutz, and Giulia Ranzini. "Privacy Cynicism: A New Approach to the Privacy Paradox." *Cyberpsychology: Journal of Psychosocial Research on Cyberspace* 10, no. 4 (December 1, 2016). https://doi.org/10.5817/CP2016-4-7.

Hope, Alexis, Ted Schwaba, and Anne Marie Piper. "Understanding Digital and Material Social Communications for Older Adults." In *SIGCHI Conference on Human Factors in Computing Systems*, 3903–12. CHI '14. New York: Association for Computing Machinery, 2014. https://doi.org/10.1145/2556288.2557133.

Hunsaker, Amanda, and Eszter Hargittai. "A Review of Internet Use among Older Adults." *New Media & Society* 20, no. 10 (July 16, 2018): 3937–54. https://doi.org/10.1177/1461444818787348.

Hunsaker, Amanda, Eszter Hargittai, and Anne Marie Piper. "Online Social Connectedness and Anxiety Among Older Adults." *International Journal of Communication* 14, no. 0 (January 28, 2020): 697–725.

Hunsaker, Amanda, Minh Hao Nguyen, Jaelle Fuchs, Teodora Djukaric, Larissa Hugentobler, and Eszter Hargittai. "'He Explained It to Me and I Also Did It Myself': How Older Adults Get Support with Their Technology Uses." *Socius* 5 (December 4, 2019): 1–13. https://doi.org/10.1177/2378023119887866.

Hunsaker, Amanda, Minh Hao Nguyen, Jaelle Fuchs, Gökçe Karaoglu, Teodora

Djukaric, and Eszter Hargittai. "Unsung Helpers: Older Adults as a Source of Digital Media Support for Their Peers." *Communication Review*, October 12, 2020, 1–22. https://doi.org/10.1080/10714421.2020.1829307.

Ingen, Erik van, and Uwe Matzat. "Inequality in Mobilizing Online Help after a Negative Life Event: The Role of Education, Digital Skills, and Capital-Enhancing Internet Use." *Information, Communication & Society* 21, no. 4 (April 3, 2018): 481–98. https://doi.org/10.1080/1369118X.2017.1293708.

Ito, Mizuko, Sonja Baumer, Matteo Bittanti, danah boyd, Rachel Cody, Becky Herr-Stephenson, Heather A. Horst, et al. *Hanging Out, Messing Around, and Geeking Out: Kids Living and Learning with New Media.* Cambridge, MA: MIT Press, 2009.

Ivan, Loredana, and Mireia Fernández-Ardèvol. "Older People and the Use of ICTs to Communicate with Children and Grandchildren." *Transnational Social Review* 7, no. 1 (January 2, 2017): 41–55. https://doi.org/10.1080/21931674.2016.1277861.

Javdan, Mohsen, Maryam Ghasemaghaei, and Mohamed Abouzahra. "Psychological Barriers of Using Wearable Devices by Seniors: A Mixed-Methods Study." *Computers in Human Behavior* 141 (April 1, 2023). https://doi.org/10.1016/j.chb.2022.107615.

Jelenchick, Lauren A., Jens C. Eickhoff, and Megan A. Moreno. "'Facebook Depression?' Social Networking Site Use and Depression in Older Adolescents." *Journal of Adolescent Health* 52, no. 1 (2013): 128–30. https://doi.org/10.1016/j.jadohealth.2012.05.008.

Jenkins, H., M. Ito, and d. boyd. *Participatory Culture in a Networked Era.* Cambridge, UK: Polity Press, 2016.

Jin, Bora, Junghwan Kim, and Lisa M. Baumgartner. "Informal Learning of Older Adults in Using Mobile Devices: A Review of the Literature." *Adult Education Quarterly* 69, no. 2 (May 1, 2019): 120–41. https://doi.org/10.1177/0741713619834726.

Kadowaki, Laura, Barbara McMillan, and Kahir Lalji. *Consultations on the Social and Economic Impacts of Ageism in Canada: "What We Heard" Report.* Federal, Provincial and Territorial (FPT) Forum of Ministers Responsible for Seniors, 2023. https://www.canada.ca/content/dam/esdc-edsc/documents/corporate/seniors/forum/reports/consultation-ageism-what-we-heard/UWBC-FinalDraftReport-WWH-EN-20230411-withISBN.pdf.

Kakulla, Brittne. "2023 Tech Trends: No End in Sight for Age 50+ Market Growth." AARP, 2023. https://doi.org/10.26419/res.00584.001.

———. *Home Tech 2020 AARP: Amerispeak Project and AAPOR Transparency Initiative Report.* Washington, DC: AARP Research, June 28, 2021. https://doi.org/10.26419/res.00420.015.

Kappeler, Kiran, Noemi Festic, and Michael Latzer. "Left Behind in the Digital Society—Growing Social Stratification of Internet Non-Use in Switzerland."

In *Media Literacy*, edited by Guido Keel and Wibke Weber, 207–24. Baden-Baden, Ger.: Nomos, 2021. https://doi.org/10.5771/9783748920656.

———. "'A Mix of Paranoia and Rebelliousness': Manifestations, Motives, and Consequences of Resistance to Digital Media." *Media Studies* 17, no. 2 (2023): 125–45. https://doi.org/10.5167/uzh-258476.

Karpinski, A. C., and A. Duberstein. "A Description of Facebook Use and Academic Performance among Undergraduate and Graduate Students." Paper presented at the Annual Meeting of the American Educational Research Association, April 13–17, 2009.

Kazlauskas, Jasmine. "Cruel Way Scammers Stole Dad's Life Savings." News.Com.Au, March 17, 2024, sec. Banking. https://www.news.com.au/finance/business/banking/cruel-way-scammers-stole-dads-life-savings/news-story/8438a9b7c07e9fa905d5fecc82dace4a.

Kelly, Heather. "For Seniors Using Tech to Age in Place, Surveillance Can Be the Price of Independence." *Washington Post*, November 19, 2021, sec. Your Data and Privacy. https://www.washingtonpost.com/technology/2021/11/19/seniors-smart-home-privacy/.

Kezer, Murat, Barış Sevi, Zeynep Cemalcilar, and Lemi Baruh. "Age Differences in Privacy Attitudes, Literacy and Privacy Management on Facebook." *Cyberpsychology: Journal of Psychosocial Research on Cyberspace* 10, no. 1 (May 1, 2016). https://doi.org/10.5817/CP2016-1-2.

Kim, Bang Hyun, and Karen Glanz. "Text Messaging to Motivate Walking in Older African Americans: A Randomized Controlled Trial." *American Journal of Preventive Medicine* 44, no. 1 (January 2013): 71–75. https://doi.org/10.1016/j.amepre.2012.09.050.

Kim, Jinsook, and Jennifer Gray. "Qualitative Evaluation of an Intervention Program for Sustained Internet Use Among Low-Income Older Adults." *Ageing International* 41, no. 3 (September 1, 2016): 240–53. https://doi.org/10.1007/s12126-015-9235-1.

Kisekka, Victoria, Rajarshi Chakraborty, Sharmistha Bagchi-Sen, and H. Raghav Rao. "Investigating Factors Influencing Web-Browsing Safety Efficacy (WSE) Among Older Adults." *Journal of Information Privacy and Security* 11, no. 3 (July 3, 2015): 158–73. https://doi.org/10.1080/15536548.2015.1073534.

"Kitboga." YouTube. Accessed April 20, 2024. https://www.youtube.com/@KitbogaShow.

Knight, Thora, Xiaojun Yuan, and DeeDee Bennett Gayle. "Illuminating Privacy and Security Concerns in Older Adults' Technology Adoption." *Work, Aging and Retirement* 10, no. 1 (January 2024): 57–60. https://doi.org/10.1093/workar/waac032.

König, Ronny, Alexander Seifert, and Michael Doh. "Internet Use among Older Europeans: An Analysis Based on SHARE Data." *Universal Access in the Infor-*

mation Society 17, no. 3 (January 19, 2018): 621–33. https://doi.org/10.1007/s10209-018-0609-5.

Korpela, Viivi, Laura Pajula, and Riitta Hänninen. "Older Adults Learning Digital Skills Together: Peer Tutors' Perspectives on Non-Formal Digital Support." *Media and Communication* 11, no. 3 (2023): 53–62. https://doi.org/10.17645/mac.v11i3.6742.

Kraut, Robert, Vicki Lundmark, Michael Patterson, Sara Kiesler, Tridas Mukhopadhyay, and William Scherlis. "Internet Paradox: A Social Technology That Reduces Social Involvement and Psychological Well-Being?" *American Psychologist* 53, no. 9 (1998): 1017–31. https://doi.org/10.1037/0003-066X.53.9.1017.

Kropczynski, Jess, Zaina Aljallad, Nathan Jeffrey Elrod, Heather Lipford, and Pamela J. Wisniewski. "Towards Building Community Collective Efficacy for Managing Digital Privacy and Security within Older Adult Communities." *Proceedings of the ACM on Human-Computer Interaction* 4, no. CSCW3 (January 5, 2021): 1–27. https://doi.org/10.1145/3432954.

Kross, Sean, Eszter Hargittai, and Elissa M. Redmiles. "Characterizing the Online Learning Landscape: What and How People Learn Online." *Proceedings of the ACM on Human-Computer Interaction* 5, no. CSCW1 (April 22, 2021): 1–19. https://doi.org/10.1145/3449220.

Last Week Tonight with John Oliver. Season 11, episode 2, "Pig Butchering Scams." Aired February 25, 2024, on HBO. Available on YouTube. https://www.youtube.com/watch?v=pLPpl2ISKTg.

Lau, Kam Man, Wai Kai Hou, Brian J. Hall, Daphna Canetti, Sin Man Ng, Agnes Iok Fong Lam, and Stevan E. Hobfoll. "Social Media and Mental Health in Democracy Movement in Hong Kong: A Population-Based Study." *Computers in Human Behavior* 64 (November 1, 2016): 656–62. https://doi.org/10.1016/j.chb.2016.07.028.

Lawrence-Lightfoot, S. *The Third Chapter: Passion, Risk, and Adventure in the 25 Years After 50*. New York: Farrar, Straus and Giroux, 2009. https://books.google.ch/books?id=PpqgmAEACAAJ.

Lee, Bob, Yiwei Chen, and Lynne Hewitt. "Age Differences in Constraints Encountered by Seniors in Their Use of Computers and the Internet." In "Group Awareness in CSCL Environments," edited by Daniel Bodemer and Jessica Dehler. Special issue, *Computers in Human Behavior*, 27, no. 3 (May 1, 2011): 1231–37. https://doi.org/10.1016/j.chb.2011.01.003.

Lee, Richard M., and Steven B. Robbins. "Measuring Belongingness: The Social Connectedness and the Social Assurance Scales." *Journal of Counseling Psychology* 42, no. 2 (1995): 232.

Leen, Eline A. E., and Frieder R. Lang. "Motivation of Computer Based Learning across Adulthood." *Computers in Human Behavior* 29, no. 3 (May 1, 2013): 975–83. https://doi.org/10.1016/j.chb.2012.12.025.

Li, Mutong. "Factors Influencing Digital Disconnection among the Elderly," 580–85. London: Francis Academic Press, 2021. https://doi.org/10.25236/iceesr.2021.098.

Linden, Sander van der. "Misinformation: Susceptibility, Spread, and Interventions to Immunize the Public." *Nature Medicine* 28, no. 3 (March 2022): 460–67. https://doi.org/10.1038/s41591-022-01713-6.

Livingstone, S. *Children and the Internet.* Cambridge, UK: Polity Press, 2009.

Livingstone, Sonia, Magdalena Bober, and Ellen Helsper. *Internet Literacy among Children and Young People: Findings from the UK Children Go Online Project.* London: UK Children Go Online, 2005. http://personal.lse.ac.uk/bober/UKCGOonlineLiteracy.pdf.

Livingstone, Sonia, and Ellen Helsper. "Gradations in Digital Inclusion: Children, Young People and the Digital Divide." *New Media & Society* 9, no. 4 (2007): 671–96. https://doi.org/10.1177/1461444807080335.

Loomba, Sahil, Alexandre de Figueiredo, Simon J. Piatek, Kristen de Graaf, and Heidi J. Larson. "Measuring the Impact of COVID-19 Vaccine Misinformation on Vaccination Intent in the UK and USA." *Nature Human Behaviour* 5, no. 3 (March 2021): 337–48. https://doi.org/10.1038/s41562-021-01056-1.

Lüders, Marika, and Petter Bae Brandtzæg. "'My Children Tell Me It's so Simple': A Mixed-Methods Approach to Understand Older Non-Users' Perceptions of Social Networking Sites." *New Media & Society* 19, no. 2 (2017): 181–98. https://doi.org/10.1177/1461444814554064.

Lutz, Christoph, Christian Pieter Hoffmann, and Giulia Ranzini. "Data Capitalism and the User: An Exploration of Privacy Cynicism in Germany." *New Media & Society* 22, no. 7 (July 1, 2020): 1168–87. https://doi.org/10.1177/1461444820912544.

Ma, Lihua, Ling Xu, Zhirui Chen, and Xueyan Zhang. "Factors of Web-Based Learning Competence among Urban Chinese Older Adults: Age Differences." *Educational Gerontology* 48, no. 5 (May 4, 2022): 210–23. https://doi.org/10.1080/03601277.2022.2031577.

Marler, Will, and Eszter Hargittai. "Division of Digital Labor: Partner Support for Technology Use among Older Adults." *New Media & Society* 26, no. 2 (January 27, 2022): 978–94. https://doi.org/10.1177/14614448211068437.

Marnfeldt, Kelly, Sindy Lomeli, Sheila Salinas Navarro, Lilly Estenson, and Kate Wilber. "'Connect It Down to the Person': Perspectives on Technology Adoption from Older Angelenos." *Journal of Elder Policy* 2, no. 3 (2023): 93–126. https://doi.org/doi:10.18278/jep.2.3.4.

Marwick, Alice, and Eszter Hargittai. "Nothing to Hide, Nothing to Lose? Incentives and Disincentives to Sharing Information with Institutions Online." *Information, Communication & Society*, March 29, 2018, 1–17. https://doi.org/10.1080/1369118X.2018.1450432.

McDougall, Kelly, and Helen Barrie. *Retired Not Expired.* Adelaide: University of Adelaide, 2020. https://www.sahealth.sa.gov.au/wps/wcm/connect/d7ef0d6e-9157-499d-a373-e3c7d249f685/Retired+Not+Expired+Report.pdf?MOD=AJPERES&CACHEID=ROOTWORKSPACE-d7ef0d6e-9157-499d-a373-e3c7d249f685-nwKW63c.

McDowell, Ian. *Measuring Health: A Guide to Rating Scales and Questionnaires*. 3rd ed. New York: Oxford University Press, 2006.

McGuire, Sandra L., Diane A. Klein, and Shu-Li Chen. "Ageism Revisited: A Study Measuring Ageism in East Tennessee, USA." *Nursing & Health Sciences* 10, no. 1 (2008): 11–16. https://doi.org/10.1111/j.1442-2018.2007.00336.x.

McNeill, Andrew, Pam Briggs, Jake Pywell, and Lynne Coventry. "Functional Privacy Concerns of Older Adults about Pervasive Health-Monitoring Systems." In *Proceedings of the 10th International Conference on PErvasive Technologies Related to Assistive Environments*, 96–102. Rhodes, Greece: Association for Computing Machinery, 2017. https://doi.org/10.1145/3056540.3056559.

Mehta, Vikram, Daniel Gooch, Arosha Bandara, Blaine Price, and Bashar Nuseibeh. "Privacy Care: A Tangible Interaction Framework for Privacy Management." *ACM Transactions on Internet Technology* 21, no. 1 (February 2021): 1–32. https://doi.org/10.1145/3430506.

Mendel, Tamir. "Social Help: Developing Methods to Support Older Adults in Mobile Privacy and Security." In *Adjunct Proceedings of the 2019 ACM International Joint Conference on Pervasive and Ubiquitous Computing and Proceedings of the 2019 ACM International Symposium on Wearable Computers*, 383–87. London: Association for Computing Machinery, 2019. https://doi.org/10.1145/3341162.3349311.

Mendel, Tamir, Debin Gao, David Lo, and Eran Toch. "An Exploratory Study of Social Support Systems to Help Older Adults in Managing Mobile Safety." In *Proceedings of the 23rd International Conference on Mobile Human-Computer Interaction*, 1–13. Toulouse, Fr.: Association for Computing Machinery, 2021. https://doi.org/10.1145/3447526.3472047.

Mendel, Tamir, Roei Schuster, Eran Tromer, and Eran Toch. "Toward Proactive Support for Older Adults: Predicting the Right Moment for Providing Mobile Safety Help." *Proceedings of the ACM on Interactive, Mobile, Wearable and Ubiquitous Technologies* 6, no. 1 (March 29, 2022): 1–25. https://doi.org/10.1145/3517249.

Mendel, Tamir, and Eran Toch. "My Mom Was Getting This Popup." *Proceedings of the ACM on Interactive, Mobile, Wearable and Ubiquitous Technologies* 3, no. 4 (December 11, 2019): 1–20. https://dl.acm.org/doi/abs/10.1145/3369821.

Mentis, Helena M., Galina Madjaroff, and Aaron K. Massey. "Upside and Downside Risk in Online Security for Older Adults with Mild Cognitive Impairment." In *CHI '19: Proceedings of the 2019 CHI Conference on Human Factors in Computing Systems*, 1–13. Glasgow: Association for Computing Machinery, 2019. https://doi.org/10.1145/3290605.3300573.

Meshi, Dar, Shelia R. Cotten, and Andrew R. Bender. "Problematic Social Media Use and Perceived Social Isolation in Older Adults: A Cross-Sectional Study." *Gerontology* 66, no. 2 (2020): 160–68. https://doi.org/10.1159/000502577.

Micheli, Marina, Elissa M. Redmiles, and Eszter Hargittai. "Help Wanted: Young Adults' Sources of Support for Questions about Digital Media." *Information, Communication & Society* 23, no. 11 (September 18, 2020): 1655–72. https://doi.org/10.1080/1369118X.2019.1602666.

Minichiello, Victor, Jan Browne, and Hal Kendig. "Perceptions and Consequences of Ageism: Views of Older People." *Ageing & Society* 20, no. 3 (May 2000): 253–78. https://doi.org/10.1017/S0144686X99007710.

Miyawaki, Christina E. "Association of Social Isolation and Health across Different Racial and Ethnic Groups of Older Americans." *Ageing & Society* 35, no. 10 (November 2015): 2201–28. https://doi.org/10.1017/S0144686X14000890.

Moore, Ryan C., and Jeffrey T. Hancock. "A Digital Media Literacy Intervention for Older Adults Improves Resilience to Fake News." *Scientific Reports* 12, no. 6008 (April 9, 2022): 1–9. https://doi.org/10.1038/s41598-022-08437-0.

Mossberger, Karen, Caroline J. Tolbert, and Mary Stansbury. *Virtual Inequality: Beyond the Digital Divide*. Washington DC: Georgetown University Press, 2003.

Müller, Andre Matthias, Selina Khoo, and Tony Morris. "Text Messaging for Exercise Promotion in Older Adults from an Upper-Middle-Income Country: Randomized Controlled Trial." *Journal of Medical Internet Research* 18, no. 1 (January 7, 2016): e5. https://doi.org/10.2196/jmir.5235.

Munger, Kevin. "Temporal Validity as Meta-Science." *Research & Politics* 10, no. 3 (July 1, 2023): 20531680231187271. https://doi.org/10.1177/20531680231187271.

Munyaka, Imani, Eszter Hargittai, and Elissa Redmiles. "The Misinformation Paradox: Older Adults Are Cynical about News Media, but Engage with It Anyway." *Journal of Online Trust and Safety* 1, no. 4 (September 20, 2022). https://doi.org/10.54501/jots.v1i4.62.

Murthy, Savanthi, Karthik S. Bhat, Sauvik Das, and Neha Kumar. "Individually Vulnerable, Collectively Safe: The Security and Privacy Practices of Households with Older Adults." *Proceedings of the ACM on Human-Computer Interaction* 5, no. CSCW1 (April 13, 2021): 1–24. https://doi.org/10.1145/3449212.

Newgate Research. *State of the (Older) Nation 2018: A Nationally Representative Survey Prepared for the COTA Federation (Councils on the Ageing)*. Sydney: Newgate Research, 2018. https://cnpea.ca/images/cota-state-of-the-older-nation-report-2018-final-online.pdf.

Ng, Reuben, and Nicole Indran. "Not Too Old for TikTok: How Older Adults Are Reframing Aging." *Gerontologist* 62, no. 8 (October 1, 2022): 1207–16. https://doi.org/10.1093/geront/gnac055.

Nguyen, Minh Hao, Eszter Hargittai, Jaelle Fuchs, Teodora Djukaric, and Amanda Hunsaker. "Trading Spaces: How and Why Older Adults Disconnect from and Switch between Digital Media." *Information Society*, August 23, 2021, 1–13. https://doi.org/10.1080/01972243.2021.1960659.

Nguyen, Minh Hao, Amanda Hunsaker, and Eszter Hargittai. "Older Adults' Online Social Engagement and Social Capital: The Moderating Role of Internet Skills." *Information, Communication & Society*, August 25, 2020, 1–17. https://doi.org/10.1080/1369118X.2020.1804980.

Nguyen, Toan, Carol Irizarry, Rob Garrett, and Andrew Downing. "Access to Mobile Communications by Older People." *Australasian Journal on Ageing* 34, no. 2 (2015): E7–12. https://doi.org/10.1111/ajag.12149.

Nicholson, James, Lynne Coventry, and Pamela Briggs. "'If It's Important It Will Be a Headline': Cybersecurity Information Seeking in Older Adults." In *CHI '19: Proceedings of the 2019 CHI Conference on Human Factors in Computing Systems*, 1–11. Glasgow: Association for Computing Machinery, 2019. https://doi.org/10.1145/3290605.3300579.

Nightingale, Sophie J., Kimberley A. Wade, and Derrick G. Watson. "Can People Identify Original and Manipulated Photos of Real-World Scenes?" *Cognitive Research: Principles and Implications* 2, no. 1 (July 18, 2017): 30. https://doi.org/10.1186/s41235-017-0067-2.

———. "Investigating Age-Related Differences in Ability to Distinguish between Original and Manipulated Images." *Psychology and Aging* 37, no. 3 (2022): 326–37. https://doi.org/10.1037/pag0000682.

Nimrod, Galit. "Aging Well in the Digital Age: Technology in Processes of Selective Optimization with Compensation." *Journals of Gerontology: Series B* 75, no. 9 (October 16, 2020): 2008–17. https://doi.org/10.1093/geronb/gbz111.

Norris, Pippa. *Digital Divide: Civic Engagement, Information Poverty and the Internet in Democratic Societies*. New York: Cambridge University Press, 2001.

Northwestern University. "Newt Minow Leads His First OLLI Study Group: School of Professional Studies, Northwestern University." March 14, 2022. https://sps.northwestern.edu/stories/news-stories/olli_newt-minow-leads-study-group.php.

Office for Seniors. *Attitudes towards Ageing: Research Commissioned by the Office for Seniors*. Wellington, NZ: Office for Seniors, 2016. https://officeforseniors.govt.nz/assets/documents/our-work/Ageing-research/Attitudes-towards-ageing-summary-report-2016.pdf.

Oliveira, Daniela, Melis Muradoglu, Devon Weir Adam Soliman, and Tian Lin Natalie Ebner. "Dissecting Spear Phishing Emails for Older vs Young Adults: On the Interplay of Weapons of Influence and Life Domains in Predicting Susceptibility to Phishing." In *CHI '17: Proceedings of the 2017 CHI Conference on Human Factors in Computing Systems*, 6412–24. Denver: Association for Computing Machinery, 2017. https://doi.org/10.1145/3025453.3025831.

Ollevier, Aline, Gabriel Aguiar, Marco Palomino, and Ingeborg Sylvia Simpelaere. "How Can Technology Support Ageing in Place in Healthy Older Adults? A Systematic Review." *Public Health Reviews* 41, no. 1 (November 23, 2020): 26. https://doi.org/10.1186/s40985-020-00143-4.

Ono, Hiroshi, and Madeline Zavodny. "Gender and the Internet." *Social Science Quarterly* 84, no. 1 (2003): 111–21.

Oxford Economics. *The Longevity Economy: How People over 50 Are Driving Economic and Social Value in the US*. Washington, DC: AARP, 2016. https://www.aarp.org/content/dam/aarp/home-and-family/personal-technology/2016/09/2016-Longevity-Economy-AARP.pdf.

Oyinlola, Oluwagbemiga. "Social Media Usage among Older Adults: Insights from Nigeria." *Activities, Adaptation & Aging* 46, no. 4 (October 2, 2022): 343–73. https://doi.org/10.1080/01924788.2022.2044975.

Page, Sydney. "100-Year-Old Woman Makes Custom Jackets by Hand and Gives Them Away." *Washington Post*, December 29, 2022. https://www.washingtonpost.com/lifestyle/2022/12/27/nancy-epley-100-jackets-virginia/.

Palfrey, John, and Urs Gasser. *Born Digital: Understanding the First Generation of Digital Natives*. New York: Basic Books, 2008.

———. *The Connected Parent: An Expert Guide to Parenting in a Digital World*. New York: Hachette, 2020. https://www.hachettebookgroup.com/titles/john-palfrey/the-connected-parent/9781541618022/?lens=basic-books.

———. *Interop: The Promise and Perils of Highly Interconnected Systems*. New York: Basic Books, 2012.

Palmore, Erdman B. "Research Note: Ageism in Canada and the United States." *Journal of Cross-Cultural Gerontology* 19, no. 1 (March 2004): 41–46. https://doi.org/10.1023/B:JCCG.0000015098.62691.ab.

Pantic, Igor. "Online Social Networking and Mental Health." *Cyberpsychology, Behavior, and Social Networking* 17, no. 10 (September 5, 2014): 652–57. https://doi.org/10.1089/cyber.2014.0070.

Pasek, Josh, Eian More, and Eszter Hargittai. "Facebook and Academic Performance: Reconciling a Media Sensation with Data." *First Monday* 14, no. 5 (2009). https://doi.org/10.5210/fm.v14i5.2498.

Peek, Sebastiaan T. M., Katrien G. Luijkx, Maurice D. Rijnaard, Marianne E. Nieboer, Claire S. van der Voort, Sil Aarts, Joost van Hoof, Hubertus J.M. Vrijhoef, and Eveline J.M. Wouters. "Older Adults' Reasons for Using Technology While Aging in Place." *Gerontology* 62, no. 2 (2016): 226–37. https://doi.org/10.1159/000430949.

Pennycook, Gordon, and David G. Rand. "Who Falls for Fake News? The Roles of Bullshit Receptivity, Overclaiming, Familiarity, and Analytic Thinking."

Journal of Personality 88, no. 2 (2020): 185–200. https://doi.org/10.1111/jopy.12476.

Perdana, Arif, and Intan Azura Mokhtar. "Seniors' Adoption of Digital Devices and Virtual Event Platforms in Singapore during Covid-19." *Technology in Society* 68 (February 1, 2022): 101817. https://doi.org/10.1016/j.techsoc.2021.101817.

Pew Research Center. *Older Adults and Technology Use*. Washington, DC: Pew Research Center, 2014. http://www.pewinternet.org/2014/04/03/older-adults-and-technology-use/.

Porter, Constance Elise, and Naveen Donthu. "Using the Technology Acceptance Model to Explain How Attitudes Determine Internet Usage: The Role of Perceived Access Barriers and Demographics." *Journal of Business Research* 59, no. 9 (2006): 999–1007. https://doi.org/10.1016/j.jbusres.2006.06.003.

Poulin, Michael J., and Claudia M. Haase. "Growing to Trust: Evidence That Trust Increases and Sustains Well-Being Across the Life Span." *Social Psychological and Personality Science* 6, no. 6 (August 1, 2015): 614–21. https://doi.org/10.1177/1948550615574301.

Powers, Richard. *The Overstory: A Novel*. New York: W. W. Norton, 2018.

Proctor, Carmel, and Roger Tweed. "Measuring Eudaimonic Well-Being." In *Handbook of Eudaimonic Well-Being*, edited by Joar Vittersø, 277–94. International Handbooks of Quality-of-Life. Cham, Switz.: Springer International, 2016. https://doi.org/10.1007/978-3-319-42445-3_18.

Quan-Haase, Anabel, and Isioma Elueze. "Revisiting the Privacy Paradox: Concerns and Protection Strategies in the Social Media Experiences of Older Adults." In *Proceedings of the 9th International Conference on Social Media and Society*, 150–59. Copenhagen: Association for Computing Machinery, 2018. https://doi.org/10.1145/3217804.3217907.

Quan-Haase, Anabel, Molly-Gloria Harper, and Alice Hwang. "Digital Media Use and Social Inclusion: A Case Study of East York Older Adults." In *Vulnerable People and Digital Inclusion: Theoretical and Applied Perspectives*, edited by Panayiota Tsatsou, 189–209. Cham, Switz.: Springer International, 2022. https://doi.org/10.1007/978-3-030-94122-2_10.

Quan-Haase, Anabel, and Dennis Ho. "Online Privacy Concerns and Privacy Protection Strategies among Older Adults in East York, Canada." *Journal of the Association for Information Science and Technology* 71, no. 9 (2020): 1089–1102. https://doi.org/10.1002/asi.24364.

Quan-Haase, Anabel, Kim Martin, and Kathleen Schreurs. "Interviews with Digital Seniors: ICT Use in the Context of Everyday Life." *Information, Communication & Society* 19, no. 5 (2016): 691–707. https://doi.org/10.1080/1369118X.2016.1140217.

Quan-Haase, Anabel, Guang Ying Mo, and Barry Wellman. "Connected Seniors: How Older Adults in East York Exchange Social Support Online and Offline."

Information, Communication & Society 20, no. 7 (2017): 967–83. https://doi.org/10.1080/1369118X.2017.1305428.

Quote Investigator. "Give a Man a Fish, and You Feed Him for a Day. Teach a Man to Fish, and You Feed Him for a Lifetime." August 28, 2015. https://quoteinvestigator.com/2015/08/28/fish/.

———. "They May Forget What You Said, But They Will Never Forget How You Made Them Feel." April 6, 2014. https://quoteinvestigator.com/2014/04/06/they-feel/.

Rader, Emilee, and Rebecca Gray. "Understanding User Beliefs About Algorithmic Curation in the Facebook News Feed." In *Proceedings of the 33rd Annual ACM Conference on Human Factors in Computing Systems*, 173–82. New York: Association for Computing Machinery, 2015. https://doi.org/10.1145/2702123.2702174.

Rader, Emilee, Rick Wash, and Brandon Brooks. "Stories as Informal Lessons about Security." In *SOUPS '12: Proceedings of the Eighth Symposium on Usable Privacy and Security*, 1–17. New York: Association for Computing Machinery, 2012. https://doi.org/10.1145/2335356.2335364.

Radloff, Lenore Sawyer. "The CES-D Scale: A Self-Report Depression Scale for Research in the General Population." *Applied Psychological Measurement* 1, no. 3 (June 1, 1977): 385–401. https://doi.org/10.1177/014662167700100306.

Raina, Divya, and Geeta Balodi. "Ageism and Stereotyping of the Older Adults." *Scholars Journal of Applied Medical Sciences* 2 (January 1, 2014): 733–39.

Ray, Hirak, Flynn Wolf, Ravi Kuber, and Adam J. Aviv. "Why Older Adults (Don't) Use Password Managers," 73–90. Paper presented at the 30th USENIX Security Symposium, August 11–13, 2021. https://www.usenix.org/conference/usenixsecurity21/presentation/ray.

Redmiles, Elissa M., Sean Kross, and Michelle L. Mazurek. "How I Learned to Be Secure: A Census-Representative Survey of Security Advice Sources and Behavior." In *CCS'16: Proceedings of the 2016 ACM SIGSAC Conference on Computer and Communications Security*, 666–77. Vienna: Association for Computing Machinery, 2016. https://doi.org/10.1145/2976749.2978307.

———. *Where Is the Digital Divide? Examining the Impact of Socioeconomics on Security and Privacy Outcomes*. Technical report, Digital Repository of University of Maryland, 2016. http://drum.lib.umd.edu/bitstream/handle/1903/18867/CS-TR-5050.pdf?sequence=1&isAllowed=y.

Reinecke, Leonard, Stefan Aufenanger, Manfred E. Beutel, Michael Dreier, Oliver Quiring, Birgit Stark, Klaus Wölfling, and Kai W. Müller. "Digital Stress over the Life Span: The Effects of Communication Load and Internet Multitasking on Perceived Stress and Psychological Health Impairments in a German Probability Sample." *Media Psychology* 20, no. 1 (January 2, 2017): 90–115. https://doi.org/10.1080/15213269.2015.1121832.

Reisdorf, Bianca C., and Darja Groselj. "Internet (Non-)Use Types and Motivational Access: Implications for Digital Inequalities Research." *New Media & Society* 19, no. 8 (August 1, 2017): 1157–76. https://doi.org/10.1177/1461444815621539.

Reuter, Arlind, Thomas Scharf, and Jan Smeddinck. "Content Creation in Later Life: Reconsidering Older Adults' Digital Participation and Inclusion." *Proceedings of the ACM on Human-Computer Interaction* 4, no. CSCW3 (January 5, 2021): 1–23. https://doi.org/10.1145/3434166.

Rich, R. Bruce. "Professor Howard Gardner Discusses His Memoir, A Synthesizing Mind." ALI Social Impact Review, September 30, 2023. https://www.sir.advancedleadership.harvard.edu/articles/professor-howard-gardner-discusses-his-memoir-a-synthesizing-mind.

Rikard, R. V., Ronald W. Berkowsky, and Shelia R. Cotten. "Discontinued Information and Communication Technology Usage among Older Adults in Continuing Care Retirement Communities in the United States." *Gerontology* 64, no. 2 (2018): 188–200. https://doi.org/10.1159/000482017.

Robinson, Laura, Shelia R. Cotten, Hiroshi Ono, Anabel Quan-Haase, Gustavo Mesch, Wenhong Chen, Jeremy Schulz, Timothy M. Hale, and Michael J. Stern. "Digital Inequalities and Why They Matter." *Information, Communication & Society* 18, no. 5 (2015): 569–82. https://doi.org/10.1080/1369118X.2015.1012532.

Rondán-Cataluña, Francisco Javier, Patricio E. Ramírez-Correa, Jorge Arenas-Gaitán, Muriel Ramírez-Santana, Elizabeth E. Grandón, and Jorge Alfaro-Pérez. "Social Network Communications in Chilean Older Adults." *International Journal of Environmental Research and Public Health* 17, no. 17 (January 2020): 6078. https://doi.org/10.3390/ijerph17176078.

Roozenbeek, Jon, and Sander van der Linden. "Prebunking: A Psychological 'Vaccine' Against Misinformation." *Journalism & Mass Communication Quarterly* 98, no. 3 (2021): 651–54.

Roozenbeek, Jon, Claudia R. Schneider, Sarah Dryhurst, John Kerr, Alexandra L. J. Freeman, Gabriel Recchia, Anne Marthe van der Bles, and Sander van der Linden. "Susceptibility to Misinformation about COVID-19 around the World." *Royal Society Open Science* 7, no. 10 (2020): 201199. https://doi.org/10.1098/rsos.201199.

Ross, Michael, Igor Grossmann, and Emily Schryer. "Contrary to Psychological and Popular Opinion, There Is No Compelling Evidence That Older Adults Are Disproportionately Victimized by Consumer Fraud." *Perspectives on Psychological Science* 9, no. 4 (July 1, 2014): 427–42. https://doi.org/10.1177/1745691614535935.

Rowe, John W., and Robert L. Kahn. "Successful Aging." *Gerontologist* 37, no. 4 (August 1, 1997): 433–40. https://doi.org/10.1093/geront/37.4.433.

———. "Successful Aging 2.0: Conceptual Expansions for the 21st Century." *Journals of Gerontology: Series B* 70, no. 4 (July 1, 2015): 593–96. https://doi.org/10.1093/geronb/gbv025.

Sala, Emanuela, Alessandra Gaia, and Gabriele Cerati. "The Gray Digital Divide in Social Networking Site Use in Europe: Results From a Quantitative Study." *Social Science Computer Review* 40, no. 2 (April 1, 2022): 328–45. https://doi.org/10.1177/0894439320909507.

Sayago, Sergio, Paula Forbes, and Josep Blat. "Older People Becoming Successful ICT Learners Over Time: Challenges and Strategies Through an Ethnographical Lens." *Educational Gerontology* 39, no. 7 (July 1, 2013): 527–44. https://doi.org/10.1080/03601277.2012.703583.

Schlomann, Anna, Alexander Seifert, Susanne Zank, Christiane Woopen, and Christian Rietz. "Use of Information and Communication Technology (ICT) Devices Among the Oldest-Old: Loneliness, Anomie, and Autonomy." *Innovation in Aging* 4, no. 2 (May 1, 2020). https://doi.org/10.1093/geroni/igz050.

Schmall, Emily. "Retirees Are Losing Their Life Savings to Romance Scams. Here's What to Know." *New York Times*, February 3, 2023, sec. Business. https://www.nytimes.com/2023/02/03/business/retiree-romance-scams.html.

Schradie, Jen. "The Digital Production Gap: The Digital Divide and Web 2.0 Collide." *Poetics* 39, no. 2 (2011): 145–68. https://doi.org/10.1016/j.poetic.2011.02.003.

Schreurs, Kathleen, Anabel Quan-Haase, and Kim Martin. "Problematizing the Digital Literacy Paradox in the Context of Older Adults' ICT Use: Aging, Media Discourse, and Self-Determination." *Canadian Journal of Communication* 42, no. 2 (2017): 359–77. https://doi.org/10.22230/cjc.2017v42n2a3130.

Schroeder, Tanja, Laura Dodds, Andrew Georgiou, Heiko Gewald, and Joyce Siette. "Older Adults and New Technology: Mapping Review of the Factors Associated with Older Adults' Intention to Adopt Digital Technologies." *JMIR Aging* 6, no. 1 (May 16, 2023): e44564. https://doi.org/10.2196/44564.

Seabrook, Elizabeth M., Margaret L. Kern, and Nikki S. Rickard. "Social Networking Sites, Depression, and Anxiety: A Systematic Review." *JMIR Mental Health* 3, no. 4 (November 23, 2016). https://doi.org/10.2196/mental.5842.

Segalov, Michael. "'We Don't Hold Anything Back': Meet the Old Gays, TikTok's Most Influential Pensioners." *Observer*, November 19, 2023, sec. Technology. https://www.theguardian.com/technology/2023/nov/19/meet-the-old-gays-tiktok-influencers-pensioners.

Selwyn, Neil. "Defining the 'Digital Divide': Developing a Theoretical Understanding of Inequalities in the Information Age." Occasional Paper 49, School of Social Sciences, Cardiff University, 2002. http://ictlogy.net/bibliography/reports/projects.php?idp=348.

Selwyn, Neil, and Keri Facer. *Beyond the Digital Divide: Rethinking Digital Inclusion for the 21st Century*. Bristol: FutureLab, 2007.

Selwyn, Neil, Stephen Gorard, and John Furlong. "Whose Internet Is It Anyway? Exploring Adults' (Non)Use of the Internet in Everyday Life." *European Journal of Communication* 20, no. 1 (2005): 5–26. https://doi.org/10.1177/0267323105049631.

Selwyn, Neil, Nicola Johnson, Selena Nemorin, and Elizabeth Knight. "Going Online on Behalf of Others: An Investigation of 'Proxy' Internet Consumers." Australian Communications Consumer Action Network, 2016. https://accan.org.au/index.php.

Seo, Hyunjin. "Community-Based Intervention Research Strategies: Digital Inclusion for Marginalized Populations." In *Research Exposed*, edited by Eszter Hargittai, 245–64. New York: Columbia University Press, 2021. https://doi.org/10.7312/harg18876-013.

Seo, Hyunjin, Joseph Erba, Darcey Altschwager, and Mugur Geana. "Evidence-Based Digital Literacy Class for Older, Low-Income African-American Adults." *Journal of Applied Communication Research* 47, no. 2 (March 11, 2019): 130–52. https://doi.org/10.1080/00909882.2019.1587176.

Shaw, Aaron, and Eszter Hargittai. "The Pipeline of Online Participation Inequalities: The Case of Wikipedia Editing." *Journal of Communication* 68, no. 1 (February 1, 2018): 143–68. https://doi.org/10.1093/joc/jqx003.

Sheldon, Pavica, Mary Grace Antony, and Lynn Johnson Ware. "Baby Boomers' Use of Facebook and Instagram: Uses and Gratifications Theory and Contextual Age Indicators." *Heliyon* 7, no. 4 (April 1, 2021): e06670. https://doi.org/10.1016/j.heliyon.2021.e06670.

Sheppard, S. Andrew, Julian Turner, Jacob Thebault-Spieker, Haiyi Zhu, and Loren Terveen. "Never Too Old, Cold or Dry to Watch the Sky: A Survival Analysis of Citizen Science Volunteerism." *Proceedings of the ACM on Human-Computer Interaction* 1, no. CSCW (December 6, 2017): 1–21. https://doi.org/10.1145/3134729.

Sims, Tamara, Andrew E. Reed, and Dawn C. Carr. "Information and Communication Technology Use Is Related to Higher Well-Being Among the Oldest-Old." *Journals of Gerontology: Series B* 72, no. 5 (September 1, 2017): 761–70. https://doi.org/10.1093/geronb/gbw130.

Smith, Tom W. "Happiness: Time Trends, Seasonal Variations, Intersurvey Differences, and Other Mysteries." *Social Psychology Quarterly* 42, no. 1 (1979): 18–30. https://doi.org/10.2307/3033870.

Solove, Daniel J. "A Taxonomy of Privacy." *University of Pennsylvania Law Review* 154, no. 3 (January 2006): 477–564. https://doi.org/10.2307/40041279.

Spirlet, Thibault. "A 74-Year-Old Man Lost $50,000 Life Savings after Downloading an App to Order Peking Duck." Business Insider, 2023. https://www

.businessinsider.com/man-lost-50k-lifesavings-downloading-app-for-peking-duck-food-2023-10.

Stanton, Jeffrey M., Kathryn R. Stam, Paul Mastrangelo, and Jeffrey Jolton. "Analysis of End User Security Behaviors." *Computers & Security* 24, no. 2 (2005): 124–33. https://doi.org/10.1016/j.cose.2004.07.001.

State of Illinois. "Articles of Incorporation: John D. and Catherine T. MacArthur Foundation." State of Illinois, Office of the Secretary of State, 1970.

Sum, Shima, R. Mark Mathews, Ian Hughes, and Andrew Campbell. "Internet Use and Loneliness in Older Adults." *CyberPsychology & Behavior* 11, no. 2 (April 2008): 208–11. https://doi.org/10.1089/cpb.2007.0010.

Szabo, Agnes, Joanne Allen, Christine Stephens, and Fiona Alpass. "Longitudinal Analysis of the Relationship Between Purposes of Internet Use and Well-Being Among Older Adults." *Gerontologist*, April 23, 2018. https://doi.org/10.1093/geront/gny036.

Taha, Jessica, Sara J. Czaja, and Joseph Sharit. "Technology Training for Older Job-Seeking Adults: The Efficacy of a Program Offered through a University-Community Collaboration." *Educational Gerontology* 42, no. 4 (April 2, 2016): 276–87. https://doi.org/10.1080/03601277.2015.1109405.

Tandoc, Edson C., Patrick Ferrucci, and Margaret Duffy. "Facebook Use, Envy, and Depression among College Students: Is Facebooking Depressing?" *Computers in Human Behavior* 43 (February 2015): 139–46. https://doi.org/10.1016/j.chb.2014.10.053.

Tripodi, Francesca. *The Propagandists' Playbook: How Conservative Elites Manipulate Search and Threaten Democracy*. New Haven, CT: Yale University Press, 2022. https://yalebooks.yale.edu/9780300248944/the-propagandists-playbook.

Tsai, Hsin-yi Sandy, Ruth Shillair, and Shelia R. Cotten. "Social Support and 'Playing Around': An Examination of How Older Adults Acquire Digital Literacy with Tablet Computers." *Journal of Applied Gerontology* 36, no. 1 (January 1, 2017): 29–55. https://doi.org/10.1177/0733464815609440.

Tsai, Hsin-yi Sandy, Ruth Shillair, Shelia R. Cotten, Vicki Winstead, and Elizabeth Yost. "Getting Grandma Online: Are Tablets the Answer for Increasing Digital Inclusion for Older Adults in the U.S.?" *Educational Gerontology* 41, no. 10 (October 3, 2015): 695–709. https://doi.org/10.1080/03601277.2015.1048165.

Turner Lee, Nicol. *Digitally Invisible: How The Internet Is Creating the New Underclass*. Washington, DC: Brookings Institution Press, 2024.

Turner Lee, Nicol, James Seddon, Brooke Tanner, and Samantha Lai. "Why the Federal Government Needs to Step Up Efforts to Close the Rural Broadband Divide." Rural Broadband Equity Project, October 4, 2022. https://www.brookings.edu/research/why-the-federal-government-needs-to-step-up-their-efforts-to-close-the-rural-broadband-divide/.

Tyler, Mark, Linda De George-Walker, and Veronika Simic. "Motivation Matters: Older Adults and Information Communication Technologies." *Studies in the Education of Adults* 52, no. 2 (July 2, 2020): 175–94. https://doi.org/10.1080/02660830.2020.1731058.

United Nations. "World Population Prospects 2019." United Nations Population Division, 2019. https://population.un.org/wpp/Graphs/Probabilistic/POP/60plus/900.

———. "World Population Prospects: The 2017 Revision, Key Findings and Advance Tables." Department of Economic and Social Affairs, Population Division, 2017.

Vaidhyanathan, Siva. "Generational Myth." *Chronicle of Higher Education* 55, no. 4 (2008): B7–9.

Vaillant, George E. *Aging Well: Surprising Guideposts to a Happier Life from the Landmark Harvard Study of Adult Development.* New York: Little, Brown, 2002.

Vaillant, G. E., and K. Mukamal. "Successful Aging." *American Journal of Psychiatry* 158, no. 6 (2001): 839–47. https://doi.org/10.1176/appi.ajp.158.6.839.

Vaportzis, Eleftheria, Maria Giatsi Clausen, and Alan J. Gow. "Older Adults Perceptions of Technology and Barriers to Interacting with Tablet Computers: A Focus Group Study." *Frontiers in Psychology* 8 (2017). https://doi.org/10.3389/fpsyg.2017.01687.

Vazquez, Christian Elias, Bo Xie, Kristina Shiroma, and Neil Charness. "Individualistic Versus Collaborative Learning in an eHealth Literacy Intervention for Older Adults: Quasi-Experimental Study." *JMIR Aging* 6, no. 1 (February 9, 2023): e41809. https://doi.org/10.2196/41809.

Vertesi, Janet. "My Experiment Opting Out of Big Data Made Me Look Like a Criminal." *Time*, May 1, 2014. https://time.com/83200/privacy-internet-big-data-opt-out/.

Vijaykumar, Santosh, Yan Jin, Daniel Rogerson, Xuerong Lu, Swati Sharma, Anna Maughan, Bianca Fadel, Mariella Silva de Oliveira Costa, Claudia Pagliari, and Daniel Morris. "How Shades of Truth and Age Affect Responses to COVID-19 (Mis)Information: Randomized Survey Experiment among WhatsApp Users in UK and Brazil." *Humanities and Social Sciences Communications* 8, no. 1 (March 23, 2021): 1–12. https://doi.org/10.1057/s41599-021-00752-7.

Vosoughi, Soroush, Deb Roy, and Sinan Aral. "The Spread of True and False News Online." *Science* 359, no. 6380 (March 9, 2018): 1146–51. https://doi.org/10.1126/science.aap9559.

Vroman, Kerryellen G., Sajay Arthanat, and Catherine Lysack. "'Who over 65 Is Online?' Older Adults' Dispositions toward Information Communication Technology." *Computers in Human Behavior* 43 (February 1, 2015): 156–66. https://doi.org/10.1016/j.chb.2014.10.018.

Waldinger, Robert, and Marc Schulz. "Essay: The Lifelong Power of Close Relationships." *Wall Street Journal*, January 13, 2023, sec. Life. https://www.wsj.com/articles/the-lifelong-power-of-close-relationships-11673625450.

Wall Street Journal Staff. "Facebook's Documents About Instagram and Teens, Published." *Wall Street Journal*, September 30, 2021, sec. Tech. https://www.wsj.com/articles/facebook-documents-instagram-teens-11632953840.

Wan, Zhiyuan, Lingfeng Bao, Debin Gao, Eran Toch, Xin Xia, Tamir Mendel, and David Lo. "AppMoD: Helping Older Adults Manage Mobile Security with Online Social Help." *Proceedings of the ACM on Interactive, Mobile, Wearable and Ubiquitous Technologies* 3, no. 4 (December 11, 2019): 1–22. https://doi.org/10.1145/3369819.

Wang, Shengzhi, Khalisa Bolling, Wenlin Mao, Jennifer Reichstadt, Dilip Jeste, Ho-Cheol Kim, and Camille Nebeker. "Technology to Support Aging in Place: Older Adults' Perspectives." *Healthcare* 7, no. 2 (June 2019): 60. https://doi.org/10.3390/healthcare7020060.

Wang, Xun, and Robin A. Cohen. "Health Information Technology Use Among Adults: United States, July-December 2022." NCHS Data Brief, no. 482. Hyattsville, MD: National Center for Health Statistics, October 31, 2023. https://doi.org/10.15620/cdc:133700.

Ware, J. E., and Barbara Gandek. "Overview of the SF-36 Health Survey and the International Quality of Life Assessment (IQOLA) Project." *Journal of Clinical Epidemiology* 51, no. 11 (November 1, 1998): 903–12. https://doi.org/10.1016/S0895-4356(98)00081-X.

Ware, Patrick, Susan J. Bartlett, Guy Paré, Iphigenia Symeonidis, Cara Tannenbaum, Gillian Bartlett, Lise Poissant, and Sara Ahmed. "Using eHealth Technologies: Interests, Preferences, and Concerns of Older Adults." *Interactive Journal of Medical Research* 6, no. 1 (March 23, 2017). https://doi.org/10.2196/ijmr.4447.

Warschauer, M. "Reconceptualizing the Digital Divide." *First Monday* 7, no. 7 (2002). http://dx.doi.org/10.5210/fm.v7i7.967.

Watkins, S. Craig. *The Young and the Digital: What the Migration to Social Network Sites, Games, and Anytime, Anywhere Media Means for Our Future*. Boston: Beacon Press, 2009.

Webster, Gemma, and Frances Ryan. "Social Media by Proxy: How Older Adults Work within Their 'Social Networks' to Engage with Social Media." *Information Research* 28, no. 1 (2023): 50–77. https://doi.org/10.47989/irpaper952.

Westwood, Sue. "'It's the Not Being Seen That Is Most Tiresome': Older Women, Invisibility and Social (in)Justice." *Journal of Women & Aging* 35, no. 6 (November 2, 2023): 557–72. https://doi.org/10.1080/08952841.2023.2197658.

Wilson, Gemma, Jessica R. Gates, Santosh Vijaykumar, and Deborah J. Morgan. "Understanding Older Adults' Use of Social Technology and the Factors Influencing Use." *Ageing & Society* 43, no. 1 (January 2023): 222–45. https://doi.org/10.1017/S0144686X21000490.

Witherell, Carol, and Nel Noddings. *Stories Lives Tell: Narrative and Dialogue in Education*. New York: Teachers College Press, 1991.

World Health Organization. "Ageing." World Health Organization. Accessed May 31, 2024. https://www.who.int/health-topics/ageing#tab=tab_1.

———. "The Determinants of Health." World Health Organization, Health Impact Assessment, 2018. http://www.who.int/hia/evidence/doh/en/.

———. "Managing the COVID-19 Infodemic: Promoting Healthy Behaviours and Mitigating the Harm from Misinformation and Disinformation." World Health Organization, September 23, 2020. https://www.who.int/news/item/23-09-2020-managing-the-covid-19-infodemic-promoting-healthy-behaviours-and-mitigating-the-harm-from-misinformation-and-disinformation.

———. "Mental Disorders." World Health Organization, April 9, 2018. http://www.who.int/news-room/fact-sheets/detail/mental-disorders.

———. "Mental Health of Older Adults." World Health Organization, December 12, 2017. http://www.who.int/news-room/fact-sheets/detail/mental-health-of-older-adults.

———. "Novel Coronavirus (2019-nCoV): Situation Report, 13." World Health Organization, February 2, 2020. https://apps.who.int/iris/handle/10665/330778.

Xie, Bo, Ivan Watkins, Jen Golbeck, and Man Huang. "Understanding and Changing Older Adults' Perceptions and Learning of Social Media." *Educational Gerontology* 38, no. 4 (April 1, 2012): 282–96. https://doi.org/10.1080/03601277.2010.544580.

YouGov. "About Our Panel." YouGov. 2024. https://today.yougov.com/about/panel.

Yu, Rebecca P., Nicole B. Ellison, and Cliff Lampe. "Facebook Use and Its Role in Shaping Access to Social Benefits among Older Adults." *Journal of Broadcasting & Electronic Media* 62, no. 1 (January 2, 2018): 71–90. https://doi.org/10.1080/08838151.2017.1402905.

Yuan, Shupei, Syed A. Hussain, Kayla D. Hales, and Shelia R. Cotten. "What Do They Like? Communication Preferences and Patterns of Older Adults in the United States: The Role of Technology." *Educational Gerontology* 42, no. 3 (2016): 163–74. https://doi.org/10.1080/03601277.2015.1083392.

Zelle. "How to Spot a Pay Yourself Scam." Zelle. July 22, 2022. YouTube video. https://www.youtube.com/watch?v=8Yk9i4uL8dQ.

Zettel-Watson, Laura, and Dmitry Tsukerman. "Adoption of Online Health Management Tools among Healthy Older Adults: An Exploratory Study." *Health Informatics Journal* 22, no. 2 (June 1, 2016): 171–83. https://doi.org/10.1177/1460458214544047.

Zhang, Fan, David Kaufman, Robyn Schell, Glaucia Salgado, Erik Tiong Wee Seah, and Julija Jeremic. "Situated Learning through Intergenerational Play between Older Adults and Undergraduates." *International Journal of Educa-*

tional Technology in Higher Education 14, no. 1 (July 3, 2017): 16. https://doi.org/10.1186/s41239-017-0055-0.

Zhou, Wei, Takami Yasuda, and Shigeki Yokoi. "Supporting Senior Citizens Using the Internet in China." *Research and Practice in Technology Enhanced Learning* 2, no. 1 (March 1, 2007): 75–101. https://doi.org/10.1142/S1793206807000269.

Zhu, Yumei, Yifan Zhou, Cuihong Long, and Chengzhi Yi. "The Relationship between Internet Use and Health among Older Adults in China: The Mediating Role of Social Capital." *Healthcare* 9 (2021): 1–15.

INDEX

Page numbers in italics refer to figures and tables.

AARP (formerly American Association of Retired Persons), 14, 90, 92, 94, 101, 173
accessibility: and adoption of technologies, 37–38, 92, 100, 173–74; of broadband, 131; and cognitive decline, 24, 69; of community online services, 170; and diversity, 58; equitable, 184–85; and interoperability, 37; of learning, 158; and support, 57–59, 181; and transparency, 37–38, 92, 100
accessibility features and options, 21, *21*, 23–26, 36, 38, 183, 193
ACMDs. *See* aged care monitoring devices (ACMDs)
adoption of technologies, 15, 17–38; ability for, 18, 160; accessibility and transparency of, 37–38, 92, 100, 173–74; of authentication technologies, 73; of basic technologies, 4; and benefits of use, 37; and cognitive decline, 24, 69; and decision-making capabilities, 18; demographics and trends of, 34; of digital technologies, 2–3; and diversity, 25, 34–36; and interoperability, 30–31, 37; of life-enhancing technologies, 84; myths about, 117; of new technologies, 14, 15, 17–38, 160; and self-efficacy, 35, 38; of simpler services, 29; of smartphones, 4–5, 171–72; of social media, 30–35, 199; and social motivators, 20; support and networks for, 35–36; variation in among older adults, 32; young people, for connection with, 19, 162
advice. *See* lessons and advice
aged care monitoring devices (ACMDs), 102
ageism, and stereotypes, 37
aging: active, 6; assumptions about, 12; and changes in circumstances, 26–27; healthy, 6, 59, 144–45; and life stages, 6–7, 27, 177; productive, 6; as spectrum, 42, 57–58; successful, 5–6, 8–9, 12, 15, 99, 105, 124, 129, 175, 187, 189; and third chapter of life, 175–76; well, 6. *See also* older adults

AI (artificial intelligence), 49, 60, 73, 92–93, 119–20, 179, 186; and scams, 73
Alzheimer's disease, 71. *See also* cognitive decline
Amazon, 97–98
Amazon Prime, 40
American Association of Retired Persons. *See* AARP (formerly American Association of Retired Persons)
Angelou, Maya, 50
anxiety, 134–35, 137, 140–42, 144–47, 150, *195*, 196, 198
Apple, and corporate educational product on privacy, 103–4
Arcadia Senior Living Bowling Green (Kentucky), 37–38
artificial intelligence. *See* AI (artificial intelligence)
authentication, two-factor (2FA), 69–70, 73, 74–75, 77, 82, 101
Authenticator (app), 69–70
autonomy, 29, 86–87. *See also* independent living
awareness, 7–8, 23, 49–50, 72, 82, 92, 100, 137, 156–57, 159, 169, 182–83, 185

Bad News (game), 127
Beautiful Theorems that Changed Math (Lee), 143
Be My Eyes (app), 7
Better Explained, 158–59
bloggers, 169
Bogle, Jack, 62
Bogleheads investments forums, 62
born digital, 1–2, 42
Born Digital (Palfrey and Gasser), 8, 10
Bracken, John, 9
broadband, 24, 57, 131, 184
Brooks, Arthur, 12

cameras, as monitoring devices, 87, 102–3
careers, 13, 40, 55, 61, 70, 85, 157, 171. *See also* employment
caregivers and facilities: advice for, 15, 180–81; changing roles of, 41; for cognitive decline, 71; and data privacy, 99, 101, 105–6; and digital alerts, 103; and misinformation, 125, 126, 128–29; and security and privacy, 70–71, 85; and social media use, 37–38; support from, 39–41, 82, 99; and well-being of older adults, 144–45, 150
cell phones, 40. *See also* mobile phones; smartphones
ChatGPT, 92–93
cognitive decline: and adoption of new technologies, 24, 69; caretakers for, 71; mild cognitive impairment (MCI), 71; and misinformation, 117, 120; and scams and fraud, 64, 69–72, 79, 117, 190; and security risks, 69–72, 79; and technology use, 72; and vulnerability, 71. *See also* Alzheimer's disease; dementia; mental health
common sense, 25, 74
community centers: advice for, 181–83; and data privacy, 99–100, 104; and learning, 157–58, 160–61, 179; and safety and security, 68; and support, 55–56, 59, 100, 179, 181–84
companies: advice for, 183; and data privacy, 101–2; and investments in older adults, 190; and security, 185. *See also* technology companies

confidence: and awareness, 137; and internet use, 45; and knowledge, 82; and patience, 53; and self-efficacy, 156–57; and skills, 52, 57, 82; and success, 79; and support, 45, 49–50, 52–53; and technology use, 45, 52–53, 99
confidentiality. *See* privacy, data
Copilot (Microsoft), 179
Cortesi, Sandra, 8
Coursera, 159
COVID-19 pandemic, 9, 19, 23, 39–41, 56–57, 95–96, 163, 200; and "info-demic," 114; and misinformation, 107–12, 114–16, 127
cybersecurity, 55, 81–82

data portability, and switching technologies, 28–31, 37, 185, 189. *See also* interoperability
data privacy. *See* privacy, data
dating apps, and scams, 68
dementia, 71, 120, 135. *See also* cognitive decline
demographics, 2, 4, 36, 119–20, 139, 147, 169–70, 194; socio-, 14, 144, 149, 194
digital age, 1, 8, 10, 41, 53, 72, 88, 91–92, 96, 98–99, 106, 124, 175
digital divide, 2, 33–34, 73, 149, 184; and inequality, 5, 9, 166–67
digital literacy, 7, 120–21, 127–28, 172, 177–78
digital media and technologies. *See* technology
"digital only" experiences, 20
disinformation, 74, 110, 124, 126–27. *See also* misinformation
diversity: and accessibility, 58; and adoption of technologies, 25, 34–36; and data privacy, 93; and learning, 159–60, 163, 169; among older adults, 1, 15, 25, 35, 70, 159–60, 169, 177; and resources, 177, 178–79; and safety and security, 70, 73–74; and skills, 35, 93; and support, 43–44, 58
DoorDash, 40
DuckDuckGo (search engine), 88
Duo (app), 69–70

e-books, 43, 55, 89
education. *See* learning; teaching
Eldera (app), 143, 150
Elder Fraud Report (FBI), 64
email, 20–21
employment, 33, 72–73, 79, 126, 137, 160, 170. *See also* careers; retirement
entertainment, 7, 23, 40, 59, 140, 145, 163
Epley, Nancy, 163–64

Facebook, 6; and academic outcomes, 146; adoption of, 21, 31–33; advice about, 179, 183; and data privacy, 95, 106; and learning, 158–59, 164–65, 167; and misinformation, 111–12, 118, 122–23, 129; and security, 61–62, 65, 70, 74, 83; and support, 41, 52–54, 56–57, 182; and well-being, 131–33, 136–37, 140–42, 146–47
FaceTime, 85, 141, 180
fake news. *See* misinformation
families: connection to, 17, 19, 61, 189; and data privacy, 99, 101, 105–6; and misinformation, prebunking, 118, 126; support by, 35, 41, 46, 87, 99. *See also* grandchildren; grandparents; learning: intergenerational; spouses and partners
Federal Bureau of Investigation (FBI), 63–64, 73

Federal Communications Commission (FCC), 153–54
frauds. *See* scams and frauds
friends: connection to, 61, 189, 190; and misinformation, prebunking, 126; support by, 35, 46
From Strength to Strength (Brooks), 12

Gardner, Howard, 10–11
Gasser, Urs, 8
Gemini (Google), 179
Global North, 184–85
Google: and authentication, 70; Gemini, 179; and misinformation, 112–13, 125–26, 129
government: and data access, 29–31; and data privacy, 104–5; lawmakers, 30
Go Viral! (game), 127
grandchildren, 17, 19, 26–27, 46, 48, 51, 85, 103, 162, 164–65
grandparents, 37, 41, 97, 142–43
Griffin, Dana, 143
Grubhub, 40

Harmony Square (game), 127
Harvard Study of Adult Development, 137–38
health: and aging, 6, 59, 144–45; challenges for older adults, 137, 177; digital tools for, 86–87, 89, 102–3; and misinformation, 113–16, 119, 129–30; and technology, 81, 138, 163, 165–67, 170, 172, 187, 190; and well-being, 4, 6, 81, 94, 133–35, 137–42, 144–45, 148–51. *See also* mental health
help. *See* support
Hulu, 43

identity theft, 63, 74–75, 149
independent living, 85–89, 102–3, 105–6. *See also* autonomy
inequality, and digital divide, 5, 9, 166–67
information overload, 157
Instagram, 21, 31–34, 41, 141, 183
Institute for Learning and Retirement, 154
instructors. *See* teaching
interconnections, 19, 40–41, 54, 59, 123, 140–44, 150–51, 165, 169, 189–90
internet: and confidence, 45; and learning, 163; and misinformation, 113; relevance of, 7–8; and safety and security, 73; skills, 7, 9, 23, 34, 42, 177–78, *196*; understanding of, 81–82. *See also* social media
interoperability: accessibility of, 37; and adoption of technologies, 30–31, 37; and data portability, 28–31, 37, 185, 189; and switching technologies, 28–31, 37

jobs. *See* careers; employment
journalists, advice for, 15, 184

Kahn, Robert Louis, 6, 9
Katz, Julia, 155
Kennedy, John F., 153
Kitboga, 80

law and policy, focused investments in, 190. *See also* policy, and policymaking; policymakers
learning, 151, 153–76; accessibility of, 158, 174; and aptitude, 175; collaborative (peer), 170, 171–72, 175; and control, 106; cooperative (team-based), 170, 172–73; and curiosity, 175; and diversity, 159–60, 163, 169; and interests or hobbies, 163–65; intergenerational, 118, 170, 173; lifelong, 156, 162, 190; new technologies, 153–76; online, 163, 190;

and online content contributions, 168, *168*; online resources, knowledge of by older adults, *159*; online resources, popularity of among older adults, 167–68, *167*; positive outcomes of, 174; practical, 170, 175; settings for, 157–58; and socialization, 156–57, 162; and social media, 168; support for, 160–61, 169–70; and teaching, 78, 160, 169–70, 173; and well-being, 15, 157, 159, 175. *See also* teaching
Lee, Hyunseung, 143
lessons and advice, 15, 177–87, 189–90; for caregivers, 15, 180–81; for community centers, 181–83; for companies, 183; for journalists, 15, 184; for older adults, 2, 15, 76–78, 99, 178–79; for policymakers, 15, 184–85; for public libraries, 181–83; for senior living facilities, 181–83; for technology companies, 190; for technology developers, 15, 37, 183; for voters, 184–85. *See also* support
libraries. *See* public libraries
LinkedIn, 27, 32–33
loneliness: counteracting and reducing, 7, 141, 146–47, 151, 190; and isolation, 7, 135; and social media, 23, 139, 146–47; and well-being, 131–51

MacArthur (John D. and Catherine T.) Foundation, 9–11
MCI (mild cognitive impairment). *See* cognitive decline
memory loss. *See* cognitive decline
mental health, 107, 114, 141, 144, 146. *See also* cognitive decline
mentors and mentoring, 41, 143–44, 150
methodology, 8, 10–14, 83, 139–40, 193–201; interview study of older adults, 200–201; and online survey platform, 14; and original synthesis, 11–13; survey of adults' social media uses, 199–200, *200*; survey of older adults, 193–99; well-being measures, 195–99, *195*. *See also* surveys
Microsoft: Copilot, 179; Word, 48–49
mild cognitive impairment (MCI). *See* cognitive decline
Minow, Burt, 135–37, 153
Minow, Mary, 154–56, 162, 175
Minow, Newton, 153–58, 162, 171, 175, 177
misinformation, 6, 15, 106–30; as challenge in digital age, 124, 129; and cognitive decline, 117, 120; combating, 120, 124–25; decision-making about, 128–29; defined, 110–11; and fake news, 107–30; forms of increasing, 120; harmful consequences of, 113–14; and manipulated images, 119–20, 125; and multigenerational interactions, 118; paradox, 120, 122–24; political, 119–22, 127–28; prebunking, 80, 126–27; and propaganda, 110; risks of, 117; and scams and fraud, 69, 74, 117–19; and social media, 113, 116, 127; as threat to everyone, 118, 124; and voting, 118, 123; and vulnerability, 121. *See also* disinformation
misophonia (selective sound sensitivity syndrome), 136–37
mobile phones, *196*. *See also* cell phones; smartphones
Moebius syndrome, 135–37, 153
MrExcel, 182
Murthy, Vivek, 135

"National Strategy to Advance Social Connection," 135

Near Technology, 156
Netflix, 40, 43
Northwestern University, 9, 154–58
notifications and announcements, 26

older adults: choices and trade-offs for, 88, 105–6; concerns about technology use, 25; defined, 2, 177; and digital experiences, 194, *196*; interviews with, 200–201; lessons and advice for, 2, 15, 76–78, 99, 178–79; and life changes, 52, 138, 149; technology use, 1–15, 25, 91, 186–87, 189–90. *See also* aging
Oliver, John, 68–69
OLLI. *See* Osher Lifelong Learning Institute (OLLI)
Olshansky, S. Jay, 143
OpenCourseWare, 159
Osher Lifelong Learning Institute (OLLI), 154–57, 162, 171, 175
"Our Epidemic of Loneliness and Isolation" (Murthy), 135
Overstory, The (Powers), 13

Papa (app), 142–43, 150
parental controls, for data privacy and security, 71
partners. *See* spouses and partners
patience, 50, 53, 178
PBS. *See* Public Broadcasting Service (PBS)
peers, 72–74, 78–79, 99, 116, 178, 180–81, 190, 197; and learning, 160–61, 170–75; and smartphones, 171–72; support from, 42, 50, 54, 56, 82
personal data. *See* privacy, data; safety and security; scams and frauds
Pew Research Center, 14, 94
phishing attacks, 65, 74, 121, 126. *See also* scams and frauds
phones. *See* cell phones; mobile phones; smartphones
"Pig Butchering Scams," 68–69
Pinterest, 31–33
policy, and policymaking, 8, 29–30, 37; regarding digital inclusion, 175, 184; and law, focused investments in, 190; for learning, 176; and privacy, 104, 128, 160; for technology, 190. *See also* policymakers
policymakers: advice for, 15, 184–85; and well-being of older adults, 150. *See also* policy, and policymaking
portability of data. *See* data portability, and switching technologies
Postcrossing, 164
Powers, Richard, 13
privacy, data, 85–106; and authentication, 75, 101; awareness about, 92, 185; and caregivers, 99, 101, 105–6; and companies, 101–2; concerns about, 89–94, 102, 105, 185; and confidentiality, 13, 102; and control, 91, 101–3, 105–6; corporate educational products on, 103–4; and digital media use, 122; and diversity, 93; and families, 99, 101, 105–6; managing, 100–101; paradox, 96–99, 122; parental controls for, 71; and password management, 101; and personal data, 15, 29–31, 60, 84, 85–106; research about, 95–96; and scams, 91; and search engines, 88; and security, 24–25, 60, 77–78, 81, 84, 99, 104, 181, 189; and social media, 94–95; support for, 60, 99–101, 185; tactics for guarding, 101; and transparency, 92, 100–101. *See also* safety and security; scams and frauds
propaganda. *See* misinformation
proxy use, 52

Public Broadcasting Service (PBS), 153–54
public libraries: advice for, 181–83; and data privacy, 99–100, 104; and learning, 157–58, 179; and misinformation, 125; and safety and security, 68; support provided by, 54, 55–56, 59, 89, 100, 125, 179, 181–85

Quora, 182

Reddit, 32–33, 164, 182
research: and reflection, 191–92; and science, 145, 163; and teaching, 9
Research Network on Successful Aging, 9
retirement, 27, 54, 61, 64, 72–73, 87, 144–45. *See also* employment
Rowe, John, 6, 9

safety and security, 15, 61–84; attacks, increased, 75; and close social help, 78; and cognitive decline, 69–72, 79; and consumer protection, 185; and control, 76–77, 79; and data privacy, 24–25, 60, 77–78, 81, 84, 99, 104, 181, 189; digital, 73, 77–79; and diversity, 70, 73–74; experiences of, 83; and mitigation strategies, 75, 77; parental controls for, 71; and password managers, 76–77; prebunking and precautionary measures, 80, 82; professional support for, 55; pushing back on as major challenge, 75–77; research and scientific literature on, 73–74, 79, 84; and risks or threats, 25, 63, 70–79, 119; and support, 55, 60, 76, 78–79, 82–83; and transparency, 38; and two-factor authentication (2FA), 69–70, 73, 74–75, 77, 82; and vulnerability, 71, 76–77. *See also* cybersecurity; privacy, data; scams and frauds
scams and frauds, 1, 178, 185; and AI, 73; avoiding, 72, 75; and cognitive decline, 64, 69–72, 79, 117, 190; and data privacy, 91; and dating apps, 68; and decision-making capabilities, 190; experiences with, 65; and misinformation, 117–19; older adults as targets, 190; older women as targets, 74; popular forms of, 65–66; prebunking and precautionary measures, 80, 82, 182; prevalence of, reported by older adults, *67*; risks of, 73–74, 119, 190; safety and security for, 62–75, 80–81, 83–84; scientific literature on, 73–74; and social media, 66–68; telltale signs of, 65–66. *See also* phishing attacks; privacy, data; safety and security
science, and research, 145, 163
search engines, 49, 68, 88, 179
security. *See* safety and security
selective sound sensitivity syndrome. *See* misophonia (selective sound sensitivity syndrome)
self-efficacy: and adoption of technologies, 35, 38; and confidence, 156–57; and control, 38, 79; and support systems, 35
senior living facilities, advice for, 181–83
Shutterfly, 141
skepticism, 69, 74, 88, 118–20, 122, 126, 129, 178
Skype, 17, 26–29
smartphones: adoption of, 4–5, 171–72; ownership among different ages, 5; and peers, 171–72. *See also* cell phones; mobile phones; texts and texting

socialization, and learning, 156–57, 162
social media: adoption of, 30–35, 199; and connections, 37–38, 61; and data privacy, 94–95; education on use of, 34; experiences, *200*; feelings while using, *143*; and health information, 166; impersonations on, as scam, 66; and internet skills, 34; and learning, 168; and loneliness, 23, 139; and misinformation, 113, 116, 127; platforms, 30–34, *32*; postings by older adults, 33; and scams, 66–68; use by older adults, 31–33, *32*, *33*, 199–200, *200*; and well-being, 131, 141–43, *143*, 147. *See also* internet
social sciences, 6, 11, 139
society: equitable and just, 59; lessons and advice for, 177; technology policy and, 190
spouses and partners, 46–47, 50–52, 54, 58
Stack Overflow, 167, 182
stereotypes, 1–2, 37, 42, 46–47, 51–52, 58, 80, 112, 117, 177–78
support, 15, 39–60, *46*, *47*; and accessibility, 57–59, 181; for adoption of technologies, 35–36; and confidence, 45, 49–50, 52–53; costs of and paying for, 54–55; for data privacy, 60, 99–101, 185; and diversity, 43–44, 58; kindness and patience as key to, 50; for learning, 160–61, 169–70; and security, 55, 60, 76, 78–79, 82–83; and self-efficacy, 35; variety of forms of, 59. *See also* lessons and advice
surveys, 11, 13–14, 21–28, 31, 33–34, 44–50, 52–53, 65–66, 75, 90–94, 101, 121, 134–35, 139–40, 144–45, 158–59, 163–68, 191, 192, 193–200. *See also* methodology
Sutton, Willie, 64
teaching: and learning, 78, 160, 169–70, 173; and research, 9. *See also* learning
technology: aging successfully with, 6, 15; concerns about, 25; as life-enhancing or life-saving, 84, 89, 105; lived experience with, 13; and new opportunities, 19; older adults, use by, 1–15, 25, 91, 186–87, 189–90; playing around with, 53; policy, 190; and simplicity over complexity, 29; social, 7, 22, 123, 138–42, 144, 150–51, 169, 187; understanding, 72, 76, 81, 103
technology companies, 28–30, 36–37, 124, 185, 190. *See also* technology developers, advice for
technology developers, advice for, 15, 37, 183. *See also* technology companies
texts and texting, 20–21, 148–50. *See also* smartphones
third chapter of life, 175–76
TikTok, 31–33, 37–38, 80, 158, 166, 179, 182, 183
Tinder Swindler, The (documentary), 68
transparency: and accessibility and adoption, 37–38, 92, 100, 173–74; and data privacy, 92, 100–101; in digital domain, 100; and personal data, 101; and security, 38
Tripodi, Francesca, 125–26
Twitch (streaming platform), 80
Twitter, 32–33, 41, 70, 118–19. *See also* X (formerly Twitter), and authentication

United Nations (UN), 2, 159

value, or usefulness, 20
Vanguard Group, 62

Vertesi, Janet, 97–99
videoconferencing, 9, 162
voting and voters: advice for, 184–85; digital, 59; and misinformation, 118, 123
vulnerability: and cognitive decline, 71; and misinformation, 121; and security, 71, 76–77

Web Content Accessibility Guidelines (WCAG 2.1), 193
WebMD, 158
well-being: and digital technologies, 130, 187; emotional, 133–34; eudaemonic, 134; and health, 4, 6, 81, 94, 133–35, 137–42, 144–45, 148–51; and isolation, avoiding, 135–36, 144; and learning, 15, 157, 159, 175; and life satisfaction and happiness, 134–35, 138, 151; and loneliness, 131–51; measures, 194, 195–99, *195*; mental, 146–47; and online learning, 157; physical, 134, 148; and positive online social engagement, 141–44; psychological, 134; and self-actualization, 134; social, 133, 134–35; and social media, 131, 141–43, *143*, 147
WhatsApp, 17, 21, 28–29, 32–33, 141, 180
WHO. *See* World Health Organization (WHO)
wikiHow, 158–59
Wikipedia, 128, 158, 166–68
wired wisdom, 1–15
work. *See* careers; employment
World Health Organization (WHO), 2, 114, 120

X (formerly Twitter), and authentication, 70. *See also* Twitter

YouTube, 35, 56–57, 80, 115, 158, 166, 179, 182

Zelle (banking app), 80
Zoom, 9, 21, 35, 40–41, 43, 56, 133, 141, 143, 155–56, 162
Zooniverse, 169